WONDERS OF ACOUSTICS.

THE TARENTELLA.

WONDERS

OF

ACOUSTICS;

OR, THE

PHENOMENA OF SOUND.

FROM THE FRENCH OF RODOLPHE RADAU.

THE ENGLISH REVISED BY

ROBERT BALL, M.A.

With Illustrations.

NEW YORK:
SCRIBNER, ARMSTRONG & CO.,
SUCCESSORS TO
CHARLES SCRIBNER & CO.,
654 BROADWAY.

CONTENTS.

ACOUSTICS;

OR,

THE PHENOMENA OF SOUND.

CHAPTER I.

SOUND IN NATURE.

Noise and Musical Sound—Voices of Animals—Language of Animals—M. L—— and the Monkeys—The Sloth or Ha-ou—Singing Birds—Insects—Reptiles and Fish—Nocturnal Life in the Forests.

SOUND is movement. Repose is dumb. All sound, all noise, tells of motion; it is the invisible telegraph which Nature uses.

Sound is an appeal to sense. It cannot be understood without the attentive ear, just as light cannot be understood without the eyes which it enlightens. In voice, and word, and song it becomes the chief and dearest tie to social life. Every one knows that the blind, who hear and speak, are better off than the deaf and dumb, who have only their eyes to learn by. It is by the voice, that offspring of the air, that living beings tell most clearly their thoughts, their needs, and their desires. The voice invites, attracts, or repulses, excites or soothes, implores or curses. As speech in man's mouth, it expresses all that mind can conceive, or heart can feel. Marvellous incarnation! which lends an invisible form to thought—which carries from soul to soul passions of emotion, faith or doubt, trouble or peace. To imagine a dumb humanity is impossible.

We propose to study sound from different points of view, without, at first, discussing the exact nature of the phenomena to which it gives rise. It will be seen afterwards that these phenomena may be explained as clearly as can be desired by the theory of vibrations, and that even the rules of music arise in a large measure from a certain number of physical and physiological facts which belong to the domain of the experimental sciences. Let not the reader feel alarmed at this, however; we will touch but lightly on this side of our subject, and we will confine ourselves for the most part to a description of the results which have been obtained, without entering into a detailed proof of the laws which we shall have occasion to lay down. This book may, therefore, be read without great effort by all who wish to understand the phenomena in the midst of which our life is spent.

The sensations that the ear experiences are generally distinguished as "musical sounds" and noises. The distinction is vague; we cannot admit any essential difference between them. All noises consist of sounds of short duration, almost instantaneous, and more or less discordant. So, also, musical sounds—or, to speak more correctly, the sounds employed by musicians—are often exceedingly short in duration, and the combination in which they are placed may be perfectly discordant. Where, then, lies the limit which separates a musical sound from a noise? It is fixed by the degree of pleasure or of pain with which it impresses an organ, whose delicacy varies with different individuals.

The most striking characteristic of noise is the irregularity and abruptness of the impressions made. The rolling of a carriage on the pavement is formed of a series of discordant explosions; the noise of falling water in a fountain is also a rapid succession of jerked or unfinished sounds.

In the soft murmur of a river, in the rustling of the leaves, the transitions are less abrupt; while in other noises, such as the long moans of the wind in the chimney, the notes rise and fall by insensible degrees. In all these cases, however, we have an irregular succession of heterogeneous sounds, which follow too rapidly to allow time for musical feeling to grow, whilst the impressions which constitute musical sound are sufficiently prolonged to be distinctly recognised. In this same fact lies the difference between spoken language and songs. Usually a confused medley of sounds, which we cannot blend in a single homogeneous sensation, is also called noise. Thus, a noise is produced by pressing the palm of the hand on the keys of a piano, and striking all the notes of the scale together. It is clear from these examples that the distinction between noise and sound is only a matter of opinion, and that we may pass by a thousand gradations from the one to the other, although the distance between the two extremes is great.

The clatter made by falling blocks of wood is called "noise" by everybody, yet here is an experiment which is often made:—We take seven pieces of hard wood of the same length and breadth, but of a thickness decreasing according to a certain law. One of these dropped alone upon a plank makes a noise seemingly without a particle of music in it, but throw them down one after another regularly, in the order of their diminishing thickness, and the seven notes of the scale are perfectly heard.

The Chinese get sounds pleasant enough for a melody

by striking upon flint stones, properly chosen and suspended by threads. Many instruments used in the orchestra really produce nothing but harmonious noises, which blend with the music to sustain the rhythm. Such are the cymbals, castanettes and triangles, &c.

Inorganic nature produces only noises. The voice of the thunder, of the storm, and of the sea are but confused noises. Yet from the wind we may win most musical notes, by presenting to it an Eolian harp, whose strings can only vibrate in a certain manner.

In the animal world we meet with an infinite variety of noises, and of musical sounds; these noises and songs constitute the language of brutes. "Birds, dogs, and other animals," says the Père Mersenne, "have quite a different cry when they are angry, or complaining, or ill, to when they are happy and well; the voice is more shrill in sorrow or anger than at other times, for bile makes the voice sharp, while melancholy and phlegmatic humour render it grave, and a sanguine temperament modulates it to softness. But the voice of animals is involuntary, while that of man is free—that is to say, men speak freely, and animals cry, sing, and use their voices according to a settled law. Many say that animals are not thus restricted, urging that nothing could be more free than the song of birds like the nightingale, the goldfinch, and others; nevertheless, it must be admitted that they only sing from necessity. It may be that delight or sorrow forces them to sing, or they may be excited by some natural instinct, which leaves them no possibility of keeping silent or of ceasing their song. And when they listen to a lute, or some other harmonious sound, and sing in imitation one to another, the sounds which they imitate so strike their imagination that they cannot be silent; for their sensitive affection, being warmed by the impression on the imagination, compels the creative faculty to move the organ of the voice."

This theory of an involuntary or necessary noise is somewhat arbitrary, for it cannot be denied that many animals contrive to hold real conversations amongst themselves. We must here quote G. E. Wetzel's interesting book, called "A New Discovery of the Language of Animals, founded upon Reason and Experience." (Vienna, 1800.) The frontispiece represents a group of superior animals, with this motto, "They never lie: truth is their language." The author endeavours to prove that animals make themselves understood by combinations of sounds, which constitute the simplest language—a language full of repetitions; that they try to make themselves understood by man, and in their turn understand his language; in a word, that it would be possible to study the idioms of different animals, and from them determine the forms and the variations of their speech.

We actually find in Wetzel's book the rudiments of a dictionary of the beasts' language filling twenty pages. The author has even tried to translate into German several dialogues of dogs, cats, chickens, and other birds in illustration of his principles. He recounts a conversation composed of little abrupt cries that he overheard between some captive frogs, the purport of which was to arrange means to facilitate their escape. It may be surmised that the drift of the conversation was not altogether clear to our linguist, for the three frogs succeeded in escaping. There is no doubt that by the careful watching of animals we may come to understand their mysterious language to a certain point, and even to speak it.

Apropos of this is an amusing story from M. Jules Richard. "Going to visit an invalid friend in a military hospital," he says, "I had made the acquaintance twelve years ago of an old Government official named L——. He was a Southerner, somewhat of a boaster but brave at bottom, who swore like a heathen, and loved animals. He had grown familiar with all the cats in the hospital; and at the hour when rations are distributed, his 'Mi-aou-ou' would bring them running from the most distant part of the building, round the old soldier's porringer. I had always supposed that the cats, deceived by the perfect imitation of their mew, or accustomed as the soldiers were to the dinner hour, came mechanically to gather round their friend. 'They understand me,' insisted the old man—'they understand me perfectly. I know cat's speech and dog's speech, but monkey speech I know better than the monkeys themselves.'

"As I smiled with an expression of incredulity, 'Will you,' said M. L——, 'come with me to-morrow to the Jardin des Plantes,* and I will show you something remarkable. That's all I have to say.'

"I took good care not to miss the appointment, and M. L—— was as punctual. He led the way to the monkey-house, and no sooner had he leant upon the outer balustrade, than I heard close beside me his guttural cry—'Kirrouu! kirrikiou! courouki! courrikiou!' I tried to imitate the sounds that came from my neighbour's mouth:

"'Kirrouu!'

"Three monkeys fell into place before L——.

"'Kirrikiou!'

"Four monkeys followed their companions.

* Zoological Gardens.

"'Courouki!'

"There were twelve.

"'Courrikiou!'

"All of them were there. L——'s discourse lasted for ten minutes, during which the monkeys—ranged in several rows, seated on the ground, their front paws crossed on their knees—laughed, nodded, listened, and replied. Yes indeed, they answered, and L—— went on in fine style with his 'Kirrouu! kirrikiou! courouki! courrikiou!' We stayed for twenty minutes, and I assure you the monkeys were not tired. Suddenly L—— made a move to go: his auditors became uneasy; then as L—— left the balustrade they uttered cries of distress. We went off, but from a distance could still see the monkeys, who climbing up the wires of their cage made signs of farewell. It seemed to me that they wanted to say, 'If you do not come again, write to us at least.'"

We sometimes hear of a cats' concert. There was a time when that might have been talked of without metaphor. There used to be cats' concerts (I do not mean those held upon the tiles at midnight), pigs' concerts, bears', monkeys', donkeys', and little birds' concerts, that sang not from gaiety of heart.

This, according to the Chronicles, is what happened at Brussels in Ascension week, 1549, in honour of a miraculous image of the Virgin:—A bear played the organ. This organ was composed of twenty cats, shut in narrow boxes; their tails were tied to cords connected with the notes of the organ. Each time that the bear struck the keys, he pulled the tails of the poor cats, and forced them to mew in tune. Musical historians also speak of organs with pigs and cats together. Conrad von der Rosen, jester to the

Emperor Sigismond, succeeded, they say, in curing his master of a deep melancholy, by playing an organ of cats, arranged in scales, whose tails he pinched by striking the keys.

Father Kircher devotes one of the most curious chapters of his "Musurgie" to the voices of animals. First of all he

Fig. 1.—The Ha-ou, or Sloth.

places the "Sloth" (in Latin called *Pigritia*, or the Ha-ou animal). He gives a description of it, together with an illustration, which he professes to have received from a provincial of his order, returned from Brazil. We give it for the sake of its curiosity.

According to this account, the sloth only makes himself heard at night: his cry is a ha-ha-ha-ha-ha, running six notes up and down the scale—doh, ré, mi, fa, sol, la, sol, fa, mé, ri, doh. These notes are uttered at regular intervals, each one being separated from the following by a short pause. When the Spaniards settled in the country, they took these nocturnal cries for the singing of men in the forests. Kircher does not stint his admiration for the voice of the sloth. "If music had been invented in America," he says, "I should not hesitate to say that it was derived from the song of this creature."

But Father Kircher has other wonders in store for us. He interprets the voices of men in a most singular fashion. Those who speak with a strong, deep voice, he classes with donkeys, after Aristotle's example. The ass truly possesses a voice strong and deep enough, and he is rash, obstinate, and rude; so those who have the same voice are rash, obstinate, and rude. Father Kircher finds no difficulty in explaining the reason of this phenomenon, and he finishes by saying that the owners of bass voices are cowardly, avaricious, unbearably arrogant in prosperity, but more timid than hares in time of danger. "Such," he says, "was Caligula." Those whose voices begin their utterance in a low key, but grow shrill before they finish, are sad, morose, and passionate like oxen. A weak, shrill voice betrays an effeminate character. With those who speak fast, a low-pitched voice bespeaks strength and courage. A shrill and piercing voice is peculiar to the goat: it indicates a petulant and wantonly nature. Nevertheless, these bad natural dispositions may be overcome by education and by the will!

Of all animals, birds are the most highly gifted as to voice. To the parrot nothing is wanting for the mimicry of human speech; but this is quite mechanical, and the wonderful faculty that we admire in the parrot indicates no advantage or superiority over other animals; in repeating the words he hears, he simply proves his utter stupidity.

The starling, the blackbird, the jay and jackdaw, who all have the thick round tongue of the parrot, are more or less clever in mimicking speech. Then why do these birds remain for ever without the expression of intelligence which speech would give them? Buffon accounts for the fact by their rapid growth in infancy, and by their early separation from their parents, who do not continue the education of their children long enough to form durable and reciprocal impressions, which are the sources of intelligence.

Those birds who have the tongue forked whistle more easily than they talk. When this natural aptitude is joined with a musical memory, they learn to repeat airs. The canary, linnet, siskin, and bullfinch are noted for their readiness to learn. The parrot, on the contrary, does not learn to sing, but imitates the cries of any animals that he hears: he mews and barks as easily as he talks.

The nightingale is the true songster of our forests. By the wonderful variety of its intonations, by the deep passion of its voice, it bears the palm from all its comrades. The nightingale's song usually begins with an uncertain, timid prelude; by degrees it becomes animated, eager, and soon we hear the brilliant, thrilling notes pour forth heavenward. The full, clear warbling alternates with low murmurs, scarcely audible; the trills and rapid runs so clearly articulated, the plaintive cadences, the long-drawn notes, the passionate sighs, give place from time to time to a short silence; then the warbling begins once more, and the woods resound with the soft and stirring accents which fill the soul with sweetness. The voice of the nightingale reaches as far as the human voice: it can be heard at a distance of upwards of a mile when the air is calm, and so much the more clearly because the nightingale only sings at night, when all is silent around. In general, it is only the male who sings, but females have been known to sing as well. In captivity the nightingales sing during nine or ten months of the year; when at liberty they only begin in April, and end in June; after this month they have a hoarse cry. To make them

Fig. 2.—The Nightingale.

sing in a cage, it is necessary to treat them well, and cheer their captivity by surrounding them with foliage. Then they will sing even better than the wild nightingales. The imprisoned nightingale will vary its natural song with such passages as please it from the songs of other birds which it has heard. Musical instruments, or a melodious voice, excite and stimulate its talent; it tries to sing in unison, or to eclipse its rivals, or to drown all the noises round. Nightingales have even been seen to drop down dead in the struggle against a rival singer.

Father Kircher, in his "Musurgie," analyses the song of the nightingale at some length. "This bird," he says, "is ambitious and eager for praise; he makes as much parade of his song as a peacock of his tail. When alone, he sings simply; but no sooner is he sure of an audience than he displays with delight the treasures of his voice, and invents the most varied modulations."

Barrington has also tried to note the song of the nightingale, but, as he himself confesses, without success. The written notes executed by the most skilful flute-player do not recall the natural song.

Barrington says that the difficulty lies in the impossibility of exactly estimating the value of each note. But, though we have not succeeded yet in transcribing this wonderful song, it has sometimes been well imitated in whistling.

Buffon tells of a man who could, by his song, so charm the nightingales, that they would come to perch upon him, and suffer him to take them in his hand.

As to the compass of the nightingale's voice, it seems not to be beyond an octave. Very occasionally some shrill sounds can be heard which mount to a higher octave, but they pass like lightning, and it is by an exceptional effort that the bird reaches such a height.

It is by no means proved that the nightingale can learn to speak, though Pliny tells of one belonging to the Emperor Claudius, who spoke Greek and Latin. Father Kircher inclines to believe that this bird could be taught to imitate human speech, "but," says he, "the story that Aldrovande relates of three nightingales who told one to another during the night all that had happened in the day, at a certain hotel in Ratisbonne, has appeared fabulous to

Fig. 3.—The Hen.

many people, or at least inexplicable without some signal imposture, or help of the devil."

Fig. 4.—The Cock.

He has also arranged in notes the songs of the cuckoo, the quail, the cock, and of the hen when she is about to lay, and when she calls her little ones. We reproduce the curious plates where he gives the result of these observations, only omitting the parrot, whose natural cry is expressed by the Greek word χαιρε, which signifies "Good morning!"

It may be said of most birds that their song is a love-call. The lark is almost the only one that can be heard from spring time to winter, and that is because it alone is faithful to its love throughout the summer. The lark

Fig. 5.—The Cuckoo.

Fig. 6.—The Quail.

sings while flying; the higher it rises the louder it sings. We can hear it even when it has disappeared in the blue of the sky. Nothing is so joyous as the exquisite notes of this song.

There is a species twice the size of the ordinary lark common in Italy and the south of France. Gifted with a strong and pleasant voice, it varies its song by counterfeiting the warble of the goldfinch, the canary, and the linnet, and even the chirp of young chickens, or the cry of a cat.

The little birds whose gay song fills the woods, orchards,

gardens, and thickets during the summer, belong for the most part to the tribe of wrens. One of the most remarkable families is that of the "pewets," who imitate the song of all the other birds so as to be mistaken for them. They might be called the mocking-birds of France.

The campanero has a clear bell-like voice, which can, it is said, be heard at a distance of more than eight miles in the

Fig. 7.—The Common Lark.

region it inhabits. Each morning it raises its song, and again at noon, when the heat has silenced all its feathered colleagues, it enlivens the solitude. There comes first a piercing cry, followed by a pause, and once more a cry that ends in a silence of six or eight minutes, which is again broken by a fresh series of cries.

Among the ancients, the swan was also reckoned with the birds gifted with the power of song, but he only sang at the hour of death. This fable was long believed, and to the present day it serves as a comparison for the last effort of a dying genius. But the voice of the swan is only a kind of croak. It is however true, according to Buffon, that we can distinguish in the cries of the wild swan a kind of modulated song, composed of clarion-like notes.

The ancients had very different ideas in the matter of harmony from our own. They adored the song of the grasshopper. Anacreon dedicated an ode to it. "Happy grasshopper!" says he, "who, on the highest branches of trees moist with dew, singest like a queen, cherished by the Muses and Phœbus, who has given thee thy sweet song." Homer compares the eloquence of the old men of Troy to the song of cicadas, and a legend relates that a trial of skill between Eunomus and Ariston, two players on the cithara, was decided by a grasshopper; for one of the former's strings snapping, the gods sent a grasshopper, which, perching on the instrument, filled so well the place of the broken string, that Eunomus was proclaimed the victor. In modern times we cannot recognise music in this insect's monotonous and piercing notes.

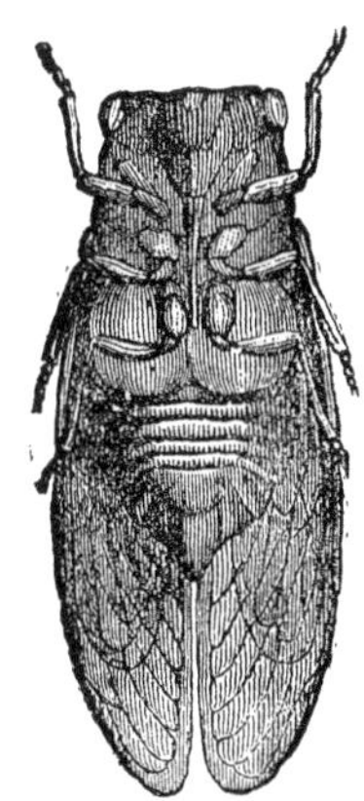
Fig. 8.— The Locust.

Its musical apparatus consists in two scaly valves placed below the abdomen, and found only in the male; these valves cover two cavities containing two membranes like dry parchment, the rapid motion of which produces a sharp, resonant, screeching noise. The other parts of the apparatus intensify and prolong the sound.

The common grasshopper is very common in Provence,

and is found also pretty far to the north ; it is met with at Fontainebleau. "In singing," says M. Maurice Gerard, "it moves its abdomen rapidly, so as to cover and uncover the openings of the sonorous cavities. Its sound is strong and sharp, and consists of one note frequently repeated, and dying away into a hiss like "st," or like air coming from a narrow aperture nearly closed. If caught it emits strong

Fig. 9.—The Hearth Cricket

cries, which differ perceptibly enough from its song when at liberty. By whistling to a grasshopper to imitate its song, you can please, attract, and easily catch it.

In northern countries the green grasshopper is often taken for the cicada, its cry being much the same. In the old editions of La Fontaine, the fable of the cicada and the ant has a grasshopper as illustration. But the two animals belong to distinct orders. Among the tribe of crickets and grasshoppers, the male calls the female by a cry produced by rubbing the elytra (wing-cases); but the mechanism that produces this monotonous noise differs a little in different species. The field-cricket rubs the whole elytra, furnished with strong, hard nerves, projecting like cords, one against the other. Travellers say that in some parts of Africa they are kept in little transparent cages: their monotonous song charms the natives to sleep

The note of the hearth-cricket is slower, more monotonous, less shrill, resembling the cry of the screech-owl.

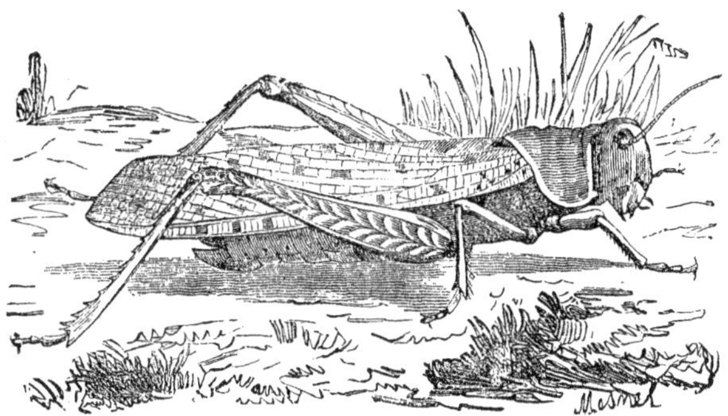

Fig. 10.—The Grasshopper.

The grasshoppers produce a cry by striking two transparent membranes, furnished with nerves placed at the base of the wing-cases, like cymbals. Their monotonous singing is heard in the evening, and all night in damp meadows. The "dectique" sings by day in the ripe wheat.

Finally, the small cricket produces sounds less musical, but more varied, than the preceding species. Their thighs and wing-cases (elytra) have hard projecting nerves, and they strike the thighs on the wing-cases, as the bow touches the chords of a violin, generally both at once, but sometimes left and right alternately. A kind of drum, covered with a very fine skin, placed near the base of the abdomen on each side of the body, seems intended to increase the sound. The cricket's cry is something like a rattle, but of different quality in different species. One can distinguish many notes, and the sound changes when calling a female, or provoking a rival.

Yersin tried to note down the song of these insects. Charles Butler, the author of the "Feminine Monarchy," tried in the same way to note the murmur of the wings that is heard in a hive of bees about to swarm. "He has fixed," says Réaumur, "all the accents of the song of the bee who aspires to lead a swarm, the different keys in which it is composed, and even the song of the queen-mother herself." The drones produce with their wings a humming noise, of which their name is an imitation.

The Death-watches, moving backwards and forwards on their six feet, strike the wood of old furniture with their closed jaws, and so cause the noise heard at night.

Reptiles are not silent. The voice of crocodiles and alligators may be compared, in infancy, to the mewing of a cat, and at a riper age to broken sobs, or bellowing, which travellers have sometimes mistaken for the cries of a child. The lizard of Birmania, M. Thomas Anquetil tells us, foretells an earthquake by its frequent and piercing cries.

Serpents have only a shrill whistle to serve as voice, excepting the rattle-snake, who carries at the end of his tail a curious instrument formed of scaly horns, fitting one into another, which become more numerous as he grows older.

The croaking frogs are renowned for their talkativeness, which, according to La Fontaine, once brought them into trouble; for, in consequence of their clamour, they were deprived of their free democracy and put under a monarchy. The fish, who pass for dumb creatures, are not so by any means. Several of them give very peculiar sounds. This power, which belongs to both male and female, is great at the time for spawning. When the "maigres" assemble in shoals, such a noise is heard to come from the water that they have gained the name of "living organs." M. Dufosse, who is specially interested in the subject, discovered that the noise is caused by the quivering of certain muscles; in some species it is sustained and strengthened by air-bladders.

Thus, by day and by night, a thousand voices join to swell the grand concert of nature. Even when we imagine ourselves in complete silence, we are still surrounded by noises. Try at such a time to listen to some very faint sound, and you will find that these noises prevent your hearing it distinctly. To feel what real silence is, one should climb the lonely summit of a high mountain. Every region has, so to say, an acoustic physiognomy. In the neighbourhood of great towns a thousand confused noises are heard, which betray human activity, as the humming of bees in a hive tells us it is inhabited.

At Paris this hoarse murmur rolls on through the night. There are streets where, in the day, a passenger cannot hear his own voice for the noise of the wheels. The rumbling is increased by the firm and elastic nature of the soil, which covers the catacombs like the sounding-board of a violin.

In Europe there are small singing birds who lead the orchestra of the forest. In America there are stronger voices to take the lead. Listen to the account Alexander von Humboldt gives of the nocturnal life, or rather the voices of the animals, in a tropical forest at night:—

He was passing the night under the spreading heavens,

having chosen a sandy plain on the banks of the Apure, bordering on a thick virgin forest. The night was cool and moonlight. A deep silence reigned on plain and river, only broken from time to time by the gentle play of the dolphins in the water. "Soon after eleven there began in the neighbouring forest such a hubbub, that any thought of sleep for the remainder of the night was out of the question. All the thicket resounded with wild cries. Amongst the many voices which mingled in this concert, the Indians could only recognise those which paused for a moment, to gather fresh vigour, and began again in a lull of the general chorus. There were the guttural and monotonous growls of the alouates, the sweet and plaintive voice of the little marmoset, the snore of the monkey, the abrupt cries of the American jaguar, of the puma, or maneless lion, of the peccary, the sloth, and a swarm of parrots. When the jaguars approached the edge of the forest our dog, who had hitherto barked incessantly, crept whimpering to find an asylum under our hammocks. Sometimes the roar of the jaguar was heard from the top of the trees, and then it was always accompanied by sharp cries of distress from the monkeys, who tried to escape this new danger."

If you ask the Indians the cause of this continued tumult, they answer, laughing, that the animals love to see the moon shine in the forest, and hold a festival at full moon. But it is not the moon which excites them most; it is during a violent storm that their cries are loudest, or when, in the midst of a peal of thunder, the lightning flashes in the forest.

These kind of scenes afford a strange contrast to the calm which reigns in the tropics towards noon in the time of the great heat, when the thermometer stands at 104° (Fahr.)

in the shade. At this time the larger animals are buried in the depths of the forest, and the birds hide themselves under the foliage of the trees, or in the crevices of the rocks, and so escape the burning rays of the sun, which pour from the zenith. To make up for this, however, the smooth rocks and stones are covered with iguanas, geckos, salamanders, who rest motionless, and with lifted head and gaping mouth seem to breathe the fiery air with delight. "But," says Humboldt, "during this apparent calm of nature, an attentive listener for almost imperceptible sounds could distinguish along the surface of the ground, in the air, a confused rustling, caused by the buzzing and humming of insects. Everything betokens a world of organic forces in motion. In each bush, in the bark torn from the trees, in the earth furrowed by the insects, life works and manifests itself. It is as one of the thousand voices of nature speaking to the thoughtful and pious soul of man."

CHAPTER II.

EFFECTS OF SOUND ON LIVING BEINGS.

Power of Music—Legends and Anecdotes—The Remedial Effects of Music—Influence of Music on Animals.

As the painter uses light for a messenger of his thought, the musician bids sound convey his feelings. Music is a language, and the sweetest of languages, inasmuch as it is less formed than any other: it is the ideal of speech.

Music is generally defined as an agreeable combination of sounds; but the ancients gave it a far wider meaning. With them music included the dance, gymnastics, poetry, and almost all the sciences. Hermes declares that music is the knowledge of the order of all things, while Pythagoras and Plato teach that everything in the universe is music. Hence the phrases "celestial music," "harmony of worlds," &c., which were used by ancient writers.

In all probability music was the first of the arts, for man had a singing master in the bird. Wind instruments must have come after. Diodorus attributes the invention of them to some shepherd, who had studied the whistling of the wind among the reeds. Lucretius holds the same opinion:

> "Et Zephyri cava per calamorum sibila primum
> Agresteis docuere cavas inflare cicutas."

Stringed instruments, and those from which sound is

produced by percussion, are also very old. The ancients attribute the invention of music to either Mercury or Apollo. Cadmus, who brought Hermione the musician to Greece,

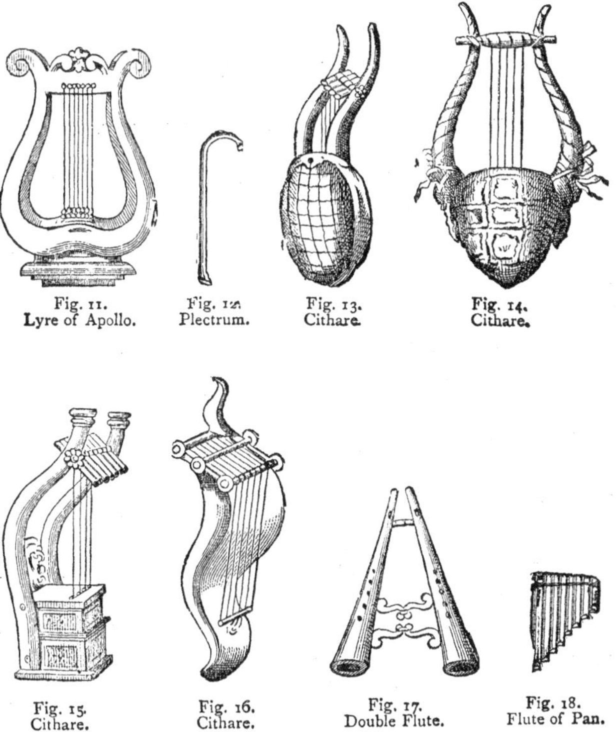

Fig. 11. Lyre of Apollo. Fig. 12. Plectrum. Fig. 13. Cithare. Fig. 14. Cithare.

Fig. 15. Cithare. Fig. 16. Cithare. Fig. 17. Double Flute. Fig. 18. Flute of Pan.

Amphion, Orpheus, and others, are spoken of as the fathers of instrumental music. According to the book of Genesis, the players on the harp and organ are descended from Jubal, the son of Lamech and Adah, of the race of Cain.

The influence of music on the manners of a people, and its power over the mind, are recognised by the philosophers of antiquity. Plato supposes that we can distinguish the sounds which incite sordid or mean feelings, as well as those which call into action the opposite virtuous feelings. With

Fig. 19.—Pastoral Pipes, or Flutes of Pan.

him it seems that a change in the popular music would be simultaneous with a change in the constitution of the state. Polybius tells us that in Arcadia, a dull and cold country, music was necessary to soften the manners of the people, and that in no place were so many crimes committed as in Cynetus, where it was neglected.

Formerly Divine and human laws, precepts and morals, legends and history, were set to music, and sung in chorus publicly. The Israelites had similar customs. Music lent a peculiar charm to abstract things, and fixed them on the mind of the hearer. Is it some memory of this sort which has recently inspired a Yankee Meyerbeer with the absurd notion of putting the American constitution into a symphony?

The Pythagoreans said that the human soul is in some way formed of harmony. They believed it possible to re-establish, by means of music, that pre-existing and primitive harmony of our intellectual faculties, too often troubled by contact with this lower world. The old writers are full of stories bearing on the miraculous power of sounds. The song of Orpheus subdued wild beasts, arrested the course of the waves, and made the trees and the rocks dance. When death had bereaved him of his Eurydice, he descended to Hades. The infernal gods, charmed by the sweetness of his music, granted him the return of his wife, whom he would have brought to earth again, if he could have abstained from looking behind during their journey. Amphion "the divine" built the walls of Thebes. At the sound of his lyre the stones came and ranged themselves one upon another:

> " ——————agitataque saxa per artem
> Sponte sua in muri membra coisse ferunt."

In the Old Testament we find music connected in a certain sense, with the destruction of a city. At the trumpet-blasts of the priests of Israel the walls of Jericho fell down.

In the songs of Finland we see the river sands change to diamonds, the haycocks run to stow themselves in barns, the sea calmed, the bears tamed by the lyre of Wainamoinen;

and he himself, falling at last under the spell, sheds in his ecstacy a torrent of pearls instead of tears.

The holy books of the Hindoos are not behind-hand in celebrating the power of music. Men and animals move in harmony with the musician's wand, while inanimate Nature obeys the influence of music composed by the god Mahédo and his wife Parbutéa. In the reign of Akbar, the celebrated singer Mia Tousine once sang a "ragà" consecrated to the night, in open day. Immediately the sun was eclipsed, and darkness spread as far as the voice was heard.* There was another "ragà" which burned him who dared to sing it. Akbar, desiring to make a trial of it, ordered a musician to sing this song while plunged up to the chin in the river Jumna. It was of no use: the unfortunate singer became a prey to the flames.

Every one knows how David played before Saul, when the evil spirit troubled the king. When Farinelli came to Spain in 1736, the accents of his voice aroused Philip V. from a deep melancholy. The king kept the musician henceforth near him, forbade him to sing in public, and loaded him with honours. He retained the same position with Ferdinand VI. This power of music on the passions has furnished material for numberless legends. They say that Alexander the Great was roused to fury by the Phrygian, and calmed by the Lydian melodies of Timotheus. There is a story too of a young man whom Pythagoras found so maddened by jealousy, wine, and a Phrygian air which had turned his head, that he was about to set fire to the house of his mistress. The philosopher of Samos simply caused

* It seems that these marvels are renewed now-a-days, for a Paris newspaper announced lately that Dreyschock had played the piano so divinely, that the wax lights shone with unwonted brilliancy.

a calmer melody to be played upon the flute, and the young maniac was brought to his senses. On another occasion, a terrible insurrection which had broken out in Lacedæmon was quelled by Terpander, who sang to the accompaniment of his harp. It might have succeeded in that age, but I doubt whether in the present day the same end could be gained by arming the police with flutes and guitars.

The Celtic priests used music for softening the manners of the people. Among the Gauls, their bards could abate the fury of combatants. St. Augustine tells how a simple flute-player excited such enthusiasm in a certain tribe that he was elected king.

There is another legend which recalls the story of Alexander the Great and Timotheus. Eric the Good, King of Denmark, heard a musician boast that he could at pleasure excite in his hearers emotions of joy, sorrow, or anger. Eric wished to put him to the proof. The musician was unwilling, and represented to the king the danger of such a trial. But the more he drew back, the more the king insisted. Seeing that it must be, the musician had all weapons removed, and arranged that some spectators should be placed outside the door, beyond the sound of his harp. They were to wait at a distance, and at a given signal to run and seize the instrument, and strike him with it. Then he shut himself up with the king and a few trusty servants, and began to play on his harp—first of all a melancholy air, which plunged the listeners in deep sadness; then changing to a joyous tone, he set them leaping and dancing. But suddenly the music became wild and fierce—they were excited beyond measure, and the king appeared in a fury. Immediately his attendants who waited outside ran, snatched the harp from the hands of the player, and struck him with it; but the king was difficult to subdue, and dealt many heavy blows before they managed to quell him under heaps of pillows. Another version tells how Eric broke open the door, seized a sword, and killed four people; of which crime he repented so bitterly that he abdicated, and afterwards set out for Jerusalem as an expiation, but died at Cyprus.

Under Henry III. the musician Claudin, playing at the wedding of the Duke de Joyeuse, excited a courtier to such a degree, that he forgot himself so far as to seize his weapons in the presence of the king; but Claudin quickly calmed him by changing the measure.

The troubadour Pierre de Chateauneuf, who lived in the thirteenth century, had a marvellous power over the feelings of his audience. Here is a story told of him by Nostradamus, in his "Lives of the Troubadours." This poet, passing through the wood of Vallongue, on his way from Roquemartine to visit the lord of the place, fell into the hands of robbers, who, after taking his money and stripping him, were about to kill him. The poet prayed them to hear a song he wished to sing before he died, and they consented. He improvised a song in praise of the brigands, and when he had finished they gave him back his horse, his money, and accoutrements, in their delight at the sweetness of his voice and verse.

A celebrated German legend tells of a wonderful magician with an enchanted flute. In the year 460 there came to Hameln, in Saxony, a man who offered to rid the town of the rats which infested it. The corporation promised him a large reward. He set himself to play upon his flute an air, which brought the rats streaming out of the houses by thousands. He drew them by his enchantment to the river Weser, where all were drowned, and he returned to claim his promised payment. But, the rats being gone, the townsfolk thought to escape their bargain, and offered him a petty sum, which he refused. He said no more, but the next day appeared with another flute, which when he played, all the children followed him. He led them to a cavern in the mountains, and they were never seen again. Then the people repented their broken faith; and since that time they date their years from "the emigration of the children," as the Turks do from the flight of the Prophet. There is a picture of the tragedy in the church at Hameln.

Without going to legends, we may find in modern history abundant notice of the power of music. Who has not heard of the "Ranz des Vaches," that air which brings home-sickness to the Swiss engaged in foreign service? At last it was forbidden, under pain of death, to play it in the army; for when they heard it the soldiers would burst into tears, or desert, or even die. "One seeks in vain," says J. J. Rousseau, "anything in this air to account for such an effect. It has no power over foreigners, and only acts on the Swiss by memory and custom—a thousand circumstances which, recalled by this music, bring to mind their native land, their old pleasures, their youth, and former ways of life, exciting sad thoughts of times gone by. The music then does not act as music, but as an aid to the memory. Although unchanged, this air has not the same power as formerly upon the Swiss, for having lost the taste for their early simplicity, they do not regret it when it is recalled. So true it is that we must not seek for the effect of music on the human heart simply in its physical action."

Military music plays an important part in the history of

battles. A quick, brilliant measure, composed of short notes, stirs the blood and incites to action. Shakespeare speaks of the "spirit-stirring" drum. How the Marsellaise has set the pulses beating!

Men are not equally sensitive to the effect of music. Some are indifferent, and some even averse to it. St. Augustine anathematises such. In his eyes a dislike for music is a sign of reprobation. This is going too far, for such an exception can only be explained by some defect in physical organisation, and one could mention many great men who suffered from this infirmity. Boyle speaks of women who were moved to tears by a tone which did not affect the rest of the audience. Rousseau mentions a lady, known to him, who could not listen to any piece of music without being seized with convulsive laughter. In the History of the Academy of Sciences we read of a musician being cured of a violent fever by a concert given in his bedroom.

It is certain that music will serve in many cases as a means of cure. Doctors of the insane often use it to calm their patients. In the Middle Ages it was believed that epilepsy, hysterics, nervous fevers, and idiotcy could be cured by music. According to Batiste Porta, a flute of hellebore cured dropsy; a flute of poplar wood, sciatica; and a pipe of cinnamon weed was a sovereign remedy for fainting fits.

Father Kircher tells us that music is the usual cure for St. Guy's dance. The sufferers in this malady dance and leap till they fall exhausted. They are cured by a strongly marked music, which excites them more and more, till it brings them to a crisis. When the disease was raging in Italy, musicians roamed the country to offer their assistance. The rapid dance they played was known by the name of

"Tarantella," a name which reminds us that the malady was supposed to be produced by the bite of the "tarantula," a large and venomous spider.

Father Kircher affirms that the spider himself has a great desire to dance when he hears that air. The experiment was tried in Andria, before the duchess and her court. They placed a tarantula on a straw, and saw him jump in time to the music.

Under the title "Phonurgia Iatrica," Father Kircher devotes a long chapter to the employment of music as a therapeutic agent. This idea should be developed, and might receive a wider application than hitherto. It is undeniable that music may be used as an exciting or calming agent, according to the rhythm of the air employed.

It is known that with children the nervous system is always excitable. The most trifling thing frightens them, or excites their imagination to great joy or sudden terror, laughter or astonishment. Their nurses quiet them by a soft lullaby. Cradled in melody, the children sleep. A joyous tune puts them in a merry mood. For this reason Montaigne always had his son awakened by music, that he might be kept in a quiet and happy temper.

Music rests or excites the mind, calms or inflames the senses, saddens or rejoices the heart. It acts even as medicine. Every one knows how a strongly-accented air helps one to walk without fatigue. The workman at the crane, and sailors at the capstan, help themselves by singing in time to their movements; and a merry, spirited waltz will set the feet tingling for a dance.

Many animals are sensitive to music, and if *all* the stories told may not be depended on, there are plenty well authenticated. At the head of these stand the singing birds, who

FIG. 20.—THE TARENTELLA (*after Kircher*).

form an orchestra of professionals. Besides these are some simple amateurs. The horse easily learns to regulate his motions to music. It is told how the Sybarites employed special musicians to train their horses to dance to the sound of flutes. One of these musicians, having a quarrel against the Sybarites, went over to the Crotonites, and excited them to war. He marched before the army with a band of musicians, and on seeing the cavalry in the distance he played familiar airs, which threw them into a confusion that ended in defeat.

It has been fancied that cattle graze more heartily to the sound of the flageolet or some other instrument, and the Arabs say that music fattens them. In the desert, when the camels are ready to drop from fatigue, the drivers encourage them with cheerful songs. Vigneul Marville, of Argonne, tells an interesting anecdote of the effects of music on different animals. While some one was playing on a marine trumpet (a kind of stringed instrument invented by Marino), he watched a cat, a dog, a horse, an ass, a doe, some cows, some birds, a cock, and some chickens which were in the court below. "The cat," he says, "seemed perfectly indifferent to the sound of the trumpet, and I judged, from her appearance, that she would have willingly exchanged all the music in the world for a mouse; she gave no sign of pleasure, but slept on in the sun. The horse stopped short under the window, and raised his head from time to time as he fed. The dog sat up on his hind legs like a monkey, with his eyes fixed on the performer; he stayed so above an hour, and seemed to delight in it. The ass gave no sign of emotion, but ate his thistles in peace. The doe pricked up her beautiful ears, and seemed very attentive. The cows stopped a little, and after having looked at us as if we were acquaintances, passed on their way. Some birds in a cage, as well as those on the trees, sang as if they would split their throats. But the cock, thinking only of his hens, and the hens, caring only for scratching and grubbing, gave me to understand that they cared nothing at all for a marine trumpet."

Buffon says that dogs are easily touched by musical sounds. "I have seen some dogs with a decided taste for music, who would come to the court-yard while a concert was going on within, and wait till the end, then return quietly home. I have seen others take the exact unison of a tone that was sounded into their ears." But there is a wide difference among dogs in this respect. Many will howl at the sound of some particular instrument, while perfectly indifferent to all others. We often see poodles show their repugnance to certain noises by twisting themselves about in the most ridiculous fashion, and howling piteously. I knew a white greyhound who always trembled when her mistress played her scales. One day, after listening silently for some time to an air that was being played, she broke out in little sharp cries, and then accompanied the piano in harmony. Surprised and pleased at this new accomplishment, her mistress fondled her, and gave her some sweetmeats. Lolette remembered the circumstance, and afterwards, whenever she had danced before the sugar cupboard in vain, she had recourse to her grand expedient and sang her song: she knew that would bring her sugar-plums. Scheitlin, in his "Psychologie Animale," asserts that dogs may be taught to pronounce certain words. I cannot tell how far this is worthy of belief.

According to Buffon, the elephant loves music, and easily learns to move in time to it, and even to join his own voice to the accompaniment of drum and trumpet. To test this theory, a concert was once given to a pair of elephants in the Jardin des Plantes. An air on the violin seemed to give much pleasure to one of them, but to the variations of the same air he was utterly indifferent. A martial air of Monsigny's had no effect on him. The thing which seemed to please him most was "Charmante Gabrielle," played upon the cornet; he listened, swinging himself on his huge legs, and grunting from time to time in unison; and occasionally he stretched out his trunk and blew, so as to nullify the sound of the cornet. When the piece was finished he fondled the musician with his trunk as if to thank him. From this account we may conclude that the elephant prefers the low notes to the high, melody to harmony, simple airs to complex, and adagio to allegro. His tastes are essentially simple.

Plutarch and Pliny add to the stock of anecdotes bearing on the same subject. We know the story of the dolphin charmed by the music of Arion—Schiller has a ballad on it. The authors of the Middle Ages believed that each animal has its favourite instrument. To the bear they allot the fife, to the stag the flute, the harp to the swan, the flageolet to the singing birds, the cymbal to the bees, and so on. Imagination evidently plays a great part in these theories. There is a more probable story of a village musician, who, returning from a wedding where he had been performing at the dance, fell into a pit in which lay a wolf. He began instinctively to scrape his violin. The wolf crouched in the opposite corner howling. He played on till the morning, frantically, madly. The strings snapped one after the other. He was at the last string, when by good fortune some villagers passed by. Their curiosity was aroused by the strange music that came from the ground. They proceeded to search out the mystery, and discovered Daniel in the ditch. He was saved, and the wolf killed.

The serpent is particularly amenable to the influence of sounds. Amongst the accounts we have of snake-charmers, who taught the serpents to dance to soft music, Chateaubriand gives us his Canadian experience in the following story :—

"In the month of June, 1796, we were travelling in Upper Canada, with some families of the tribe of Onontagués. One day, when encamped in a plain on the banks of the river Jenesie, a rattle-snake made its appearance. There was a Canadian with us who played the flute; he wished to exhibit his power, and advanced towards the animal with the novel weapon. At his approach the reptile raised itself in a spiral, flattened its head, inflated its cheeks, and drawing back its lips, displayed its poisonous fangs and cruel jaws; its forked tongue glanced like a flame, its eyes shone like coals, its body swollen with rage rose and fell like the billows of a furnace, its skin became stiff and horny, and its tail moved with such rapidity as to look like a vapour, making the while a horrid sound. Then the Canadian begins to play on his flute. The serpent draws back his head with a motion of surprise. As it falls under the magical influence, its eyes lose their awful glitter, the vibration of its tail lessens, and the noise dies away. The coils of the snake relax by degrees, taking a wider circuit, and at last they lie one by one upon the ground in concentric circles. The shades of blue and green, of white and gold re-appear in all their brilliancy on its sensitive skin, and lightly turning its head it rests motionless, in an attitude of attention and pleasure. At this moment the Canadian walks a few steps, still playing on his flute a sweet, monotonous air; the reptile lowers his neck, and dividing the fine grass with his head, crawls on in the footsteps of the musician who leads him, stopping when he stops, and following when he goes. He was thus led outside the camp, in the midst of a crowd of spectators, native and European, who could scarcely believe their eyes, and with unanimous voice it was agreed that the wonderful creature should be allowed to escape."

Lizards are also said to be remarkably alive to the influence of music. Père Labat went to a lizard-hunt with a negro armed with a noose at the end of a pole. They soon found one stretched in the sun upon the branch of a tree. The negro began to whistle to the animal, who stretched his neck to see where the sound came from. Then the negro quietly approached, still whistling, and tickled the creature's sides and throat with the end of the rod. The lizard, in delight, rolled over on his back, stretching his neck for the caress, and when within reach the noose was slipped over him.

The love of the spider for music is also well known. M. Michelet tells the following anecdote:—" Berthome, the celebrated violinist, owed his early success to the seclusion in which he was made to work while very young. But in his solitude he had one companion unsuspected—viz., a spider. First of all it lived in the corner of the wall, but gradually it ventured to the corner of the desk, then on to the child, and at last it would take up its place on the arm that held the violin, where it listened, breathless with delight and emotion. It served as an audience; the child-artist needed no other encouragement—no other sympathy But the child had a stepmother, and she one day, bringing a stranger to hear the boy's practice, saw the creature at its accustomed post, and with a single blow from her slipper annihilated the audience. The child took it so to heart that he was ill for three months, and almost died."

Whence comes the power that music exercises over the soul? What is the secret affinity by which sounds excite passions?

Music is the image of motion. It employs sounds arranged in regular intervals, between which the voice mounts and falls, according to the fancy of the musician. In varying the duration and the intensity of the different notes that succeed one another, every shade of expression, every possible difference of time is given, from the drowsy meandering of a stream, which loses itself in the sands, to the stormy impetuosity of a mountain torrent. Now sounds act directly on the nervous system by the vibration they impart to the sensitive nerves, and thus they provoke the disposition of mind agreeing to the kind of movement expressed by the music. Gaiety is characterised by a measure quick and light, gravity by a slow and solemn movement, anger by an abrupt and hasty staccato. These different characteristics apply equally well to the motions of the body, and it is in this unanimity of impression and action in soul and body that we must seek the explanation of the effects of music. Sorrow paralyses our limbs, while it makes our speech slower, and stops the flow of ideas. Music composed of notes which painfully climb a slow ascent of semitones disposes to melancholy reverie, while, on the contrary, notes which leap by fifths and octaves fill us with a flutter of excitement, which has its symbolic expression in laughter and the dance. This explanation of the psychological effects of music has not escaped Aristotle. "Why," says he, "do rhythms and melodies adapt themselves to moods of the mind, and not flavours, or colours, or odours? Is it because they are movements corresponding to actions? Their intrinsic power rests on a certain tone, and also gives this tone. Flavours and colours do not act so."

There are other movements which produce just the same effects upon us. The cascade which falls from the height of a rock, the limpid stream which ripples softly in its sandy bed, the waves that beat unceasingly on the shore, affect us like visible music. One could watch the waves for hours together break upon the level strand. "The rhythm of this movement," says Helmholtz, "which is not without a continual change in detail, awakens a feeling of repose without tedium, and generates an idea of life wide and grand, but in perfect and harmonious order. When the sea is calm we can be pleased for a time by watching its beautiful colours, but this pleasure does not last as when the water is agitated. The ripples which are found on small sheets of water are too hurried in their motions, and rather worry than soothe the spirit."

CHAPTER III.

PROPAGATION OF SOUND BY DIFFERENT MEDIA.

Effect of Vacuum—Propagation of Sound in Gas—In Water—In the Earth—Experiments by Mr. Wheatstone—Hearing by the Teeth.

How is sound carried to the ear which hears it? What is the invisible bridge by which it travels? The answer is easy. A light and elastic fluid surrounds us on every hand. The winds show us that it can produce the most powerful mechanical effects. Every commotion at once disturbs this fluid, and is felt at a considerable distance from its starting-point. Is it not, therefore, natural to admit that the aërial fluid in like manner propagates the movements which produce sound? We see, besides, that any violent explosion is always accompanied by a sudden displacement of the air. Hence scientific men have not hesitated to say that the air is the material vehicle of sound. There is no sound without air. Here is the simple experiment by the aid of which we may prove the fact:—Hang a small bell by a silk thread in a glass globe, from which the air has been exhausted by means of an air-pump. Ring the bell by shaking the thread. Nothing is heard. The tongue strikes the metal, but the blow is in a vacuum. But if the stop-cock be opened, and the air

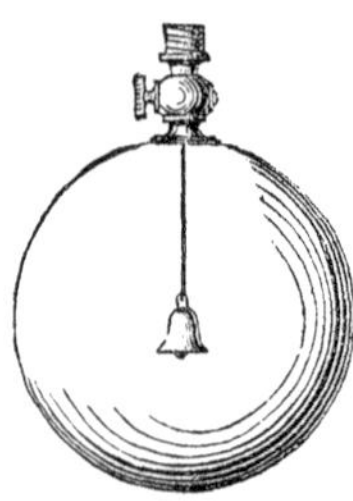

Fig. 21.

allowed to enter, the charm is broken and the bell no longer dumb (Fig. 21). This experiment may be made with an alarum introduced into the receiver of an air-pump. At first it is heard distinctly, but as the air becomes rarified the sound grows fainter, and at last dies away. You may even fire a small pistol under the receiver. You see the flash but hear no report. These experiments will only succeed when the pistol or alarum is placed upon a wadded cushion, which deadens the sound. If not, the vibration is transmitted to the stand of the air-pump, and from that to the surrounding air, which carries it to the ear. For this reason it is difficult completely to insulate the sound which is produced in the interior of the receiver. It is through having forgotten this means of communication between the sounding body and the external air, that Kircher thought he had found in the same experiment a conclusive argument against the existence of a vacuum. He had exhausted the air from a hundred feet length of leaden tubing, which terminated above in a glass chamber, in which were fixed a bell and a small hammer that could be raised from outside. When the hammer fell upon the bell, it gave out a clear ringing sound, and from this Kircher concluded that this supposed "vacuum" was but a fiction of the philosopher.

According to Kircher, thick and massive bodies such as walls or rocks do not transmit sound directly. How is it, he asks, that if one person strikes a wall, another can hear the noise by placing his ear on the opposite side? This transmission he explains by the presence of air in the pores of all bodies. It is this confined air which conducts the sound. If a body be very dense it only allows a small portion of sound to pass, because it contains but a small portion of air. He states that glass is the least porous of all substances, and that a mouse shut up in a glass chamber hermetically sealed would hear nothing in his prison, whatever noise was made outside. Kircher adds that there is in Scotland a rock called "The Deaf Rock," hiding behind which one cannot hear even the firing of a cannon. The reason of this phenomenon must be sought in the excessive density of the rock. It is, he says, opaque as to sound, just as other bodies are opaque to the light.

Although it is true that it is usually by the intervention of air that sounds reach the ear, it is now known that the presence of a gaseous fluid is not necessary to their transmission. All elastic bodies—gaseous, liquid, or solid—conduct sound. A repeater plunged in water under a glass bell may be heard distinctly above. The divers can hear under water what is going on at the surface. It is true the sound reaches them but faintly; that is because it loses intensity by entering into a medium more dense than the air. The motion which has once passed into the water is carried on there without hindrance. This is proved by the fact that sound is as distinct at the depth of several feet as close to the surface. Were it otherwise the organs of hearing in fish would be utterly useless. It is certain they can hear: tame fish have been known to respond to a whistle.

Solid bodies are good conductors of sound. The tick of a watch held at one end of a hewn trunk is heard perfectly at the other, not because of the air in the pores of the wood, but because the wood resounds under the beating of the escapement wheel. By listening with the ear on the ground, the report of a cannon can be heard at a distance of twenty-five miles, and in the same manner the trampling of horses' feet is audible from a great distance.

"Scuta sonant, pulsuque pedum tremit excita tellus."—VIRGIL.

This transmission of sound may be rendered visible by placing a drum, covered with small pebbles, on the ground. When horsemen pass, even at some distance, these pebbles are seen to move about. In the Cornish mines they excavate far out under the sea, and there, at this great depth, the noise of the waves and even the rushing murmur of the shingle can be heard. In the opposite workings of mines, too, the miners can hear through the intervening soil, and are able to direct one another. Such subterranean noises have given rise, doubtless, to many of the most thrilling ghost stories.

It appears that wood conducts sound better than any other solid body. Deal is a better conductor than box, and box than oak. With four deal rods Wheatstone contrived to carry the sounds of a concert held in the cellar through several storeys to the upper room of a house. It was done in this way:—The rods rested, one upon the sounding-board of the piano, another upon the bridge of the violin, the third upon the violoncello, and the fourth touched a clarionet. They passed through the roof of the cellar, and on to the upper storey where the audience sat. Each rod ended in a sounding-board of thin and elastic wood. The whole structure vibrated considerably when a piece of music was played below, and the room above was filled with sounds, which seemed to proceed from witchcraft. Indeed this experiment has a magical effect. The wood suddenly sings as if it were alive, and a listener, trusting to his ears alone, would fancy himself in the presence of a real orchestra. Mons. Kœnig tried the same experiment with a musical box shut in a large padded chest. A lath of wood passed out from the interior, and was surmounted by a sounding-board. When this shelf was lifted nothing could be heard, but no sooner was it placed on the free end of the lath than the tune which was being played inside the chest became perfectly audible.

The bony parts of the head act as sound-conductors to the ear. Thus sound can be communicated by the forehead or by the teeth. Two people holding a thin slit of wood between their teeth, and talking in an undertone, can hear one another at a considerable distance. The stethoscope, invented by Laennec in 1819, is based on the same principle. It consists of a wooden cylinder, which the doctor places on the chest of the patient, so as to hear more plainly the noise from the heart. Wheatstone has proposed an instrument, to which he gives the name of microphone, also intended to facilitate the hearing of very faint sounds. It is a small copper basin, which is placed over the ear, and to the centre is fixed a long metal stem, a kind of tentaculum, which carries the sound. Such an apparatus might be fitted to each ear, the stems uniting into a single tube.

If you strike on a silver spoon, a glass bell, or any other sonorous body suspended by a thread, the free end of which is introduced into an ear-trumpet, or held between the teeth, the ears being stopped, you will hear a deep, full sound, like a distant bell. A Danish physician, Herhold, tried this with a spoon fastened to a thread nearly 700 feet in length, of which one end was fastened to a pole, and the other held between his teeth.

The deaf and dumb can hear well by their teeth, when the deafness does not proceed from paralysis of the nerves. If you make them hold the edge of a musical box, or a rod of wood resting on the sounding-board of a piano, between their teeth, they will hear the sound of the instrument.

One who is partially deaf can understand easily what is said to him, if the words be spoken into a copper or glass basin which is applied to his ear or teeth.

Dull bodies, such as hemp, wadding, stuffs of all kinds, flour, and sawdust, do not sensibly transmit sounds. A Turkey carpet stifles the sound of footsteps, while a thick door-curtain prevents the sound of voices passing from room to room.

CHAPTER IV.

INTENSITY OF SOUND.

Circumstances that Vary the Intensity of Sound—Intensity at Night—Extent or Reach of Sound—The Inverse Square of Distance—Speaking Trumpets—Acoustic Tubes—Acoustic Cornets.

THE strength or intensity of sound is determined first of all by the energy of the movement which produces it, but the effect on the ear depends on the nature of the medium by which it is conducted. We have already seen that under the bell of an air-pump any sound will die away gradually, as the air becomes rarified. On high mountains, where the air has not much density, all noises lose their force, and seem more distant than they really are. At the summit of Mont Blanc, 15,000 feet above the sea-level, Saussure found that a pistol report sounded no louder than a cracker in the plains. In some experiments tried at Quito between two stations, the one at an altitude of 10,000 and the other 13,000 feet, the report of a nine-pounder cannon, fired at twelve miles distance, did not equal that of an eight-pounder heard in the plains of Paris at a distance of twenty miles. Aëronauts have often told how feeble their voices become in the high regions of the air. A railway whistle was heard at a height of three miles and a half or four miles. That is the greatest distance at which the human ear has been able to catch sounds from the earth. At this time the air was unusually damp.

In thinking of the diminution that all sound is subject to in the upper air, one is surprised at the intensity of the noise sometimes produced by an explosion of a thunderbolt. A meteor which was observed in 1719, and according to Halley's calculation travelled through the air at a height of more than sixty miles, burst with an explosion equal to that of a great cannon.

Thunderbolts generally burst with a great noise, and since we know the explosion is very high above the surface of the earth, it must be of almost inconceivable violence.

In a confined space sound is exaggerated. In the tunnels where the workmen laboured at the foundations of the Bridge of Arcueil, every sound took a metallic ring—even the voice produced an unpleasing ringing effect in the head.

Priestley made very many experiments, using different gases in place of air. Having filled a receiver with hydrogen, and put a bell within it, he found that the sound ceased almost instantly. The density of hydrogen is only one-fourteenth that of the air. Pilâtre de Rosier, having breathed great quantities of this gas, found his voice had become feeble and nasal. Mannoir and Paul did the same thing at Geneva, and their voices were singularly shrill and thin.

Sound has much greater force in water. By experiments tried on the Lake of Geneva, Colladon estimated that a bell submerged in the sea might be heard at a distance of more than sixty miles. Franklin asserts that he has heard the striking together of two stones in the water half a mile away.

When sound passes from one medium to another of a different density, it loses more or less in intensity. As before stated, divers hear noises from the surface but faintly, while those outside can hear well what passes under water. For instance, the stroke of a bell can be heard at a depth of forty feet. From this we conclude that water gives vibrations to the air more easily than air to water. If the vibrations of a solid body, instead of passing directly through the air, are conveyed through an intermediate liquid, the result is increased power. Perolle has experimented on this. He took a watch, carefully sealed with wax, and suspended it by a thread in a vase, which he filled successively with different liquids. In the air the tick of the watch became imperceptible at ten feet distant. Liquids strengthened the sound. In spirits of wine the watch was heard at thirteen feet; in oil, at sixteen; in water, at nineteen feet: by all which we see that the force of sound augments with the density of the fluid through which it passes to the air.

The vibrations of a solid body travel with difficulty through a gaseous medium; a large surface is necessary to increase the sound, and for this reason a sounding-board heightens the effect of any musical instrument connected with it, by conveying the vibration to a large mass of air, as already described in Wheatstone's experiments.

In passing through the air itself, ascending or descending, sound must cross layers, so to say, of varying density. Saussure and Schultes have stated that sound travels better up than down a mountain height, and aëronauts notice the same thing. This may be explained by the fact that the voice, and all other sounds, have, even at the moment of their production, less power in the rarified air of the higher regions than in the denser air of the plains.

When the air is unequally heated by the rays of the sun and other means, sound loses in power, and does not travel far.* By this circumstance Humboldt explains the difference in intensity of sound by night and by day. Nicholson seeks to account for it by the absence of those thousand confused noises which during the day disturb the atmosphere around us. The silence of night, he says, rests our organs, and renders them more alive to slight impressions. Silence makes our hearing more acute, as obscurity tends to sharpen our sight. But Humboldt brings his observations in America to bear against this opinion. In tropical countries the animals make more uproar at night than in the day, and the wind only rises after sunset. Yet the noise of the cataracts of Orinoco is heard at Aturés (more than a league distant) three times more plainly at night. Humboldt has also remarked that this nocturnal increase in the intensity of sound is more noticeable in the lower plains than on the table-lands or at sea.

It would, perhaps, be more correct to attribute these facts to the united influence of the different causes mentioned, to which might be added the coldness of night. It is as true indoors as in the open country, that night intensifies sound. A mouse nibbling at the wainscot sounds altogether different by night and by day. This cannot be from any inequality in the density of the air, and we must account for it by the contrast of silence. Darkness may also count for something. Many people shut their eyes in order to hear the better, and the sense of hearing is generally very acute with the blind.

We have just said that cold is favourable to the propaga-

* At the time when the ventilation of the Houses of Parliament was under discussion, it was stated that the current of heated air, which rose from the hall, prevented the voice of the speaker being heard at the opposite side.

tion of sound. This is a fact acknowledged by many. In Polar regions Captain Parry often heard a conversation, carried on in an ordinary voice, at a mile distance. One of his comrades at Port Bowen was able to converse with some of the crew 6,700 feet off, the thermometer standing at the time 28° below zero. It might be supposed that this phenomenon is due to the condensation of the air, but the experiments of Bravais and Martin do not confirm such an opinion. They ascertained that at St. Chéron a diapason mounted on a sounding-board was heard at 833 feet distance soon after midday, and at midnight the sound reached nearly 1,243 feet. On the Faulhorn the sound was heard at above 1,804 feet by night, and even on Mont Blanc at 1,105 feet, although the air is much less dense on these heights than in the plains. This unexpected result proves that it is not the condensation due to the cold which produces an increase in the intensity of sound; the phenomenon is evidently more complex, and it may probably be accounted for in some degree by the wonderful calm of mountain and Polar regions.

The wind has much to do with carrying sound. In the direction of the wind, of course, it travels far. De Haldat made some experiments near Nancy with a small drum, from which he concluded that with the wind it travels two or three times farther than against it. Since then Delaroche and Dunal have taken more exact measurements in the plain of Arcueil. They placed themselves between two drums of equal size, which were beaten with equal force, and ascertained the distance at which the two sounds seemed of the same intensity, when a straight line drawn from one drum to the other made a given angle with the direction of the wind. The faintest sound was that which came from the nearest drum. In this way they found that for distances of eighteen or nineteen feet the influence of wind was imperceptible; above that it became appreciable, and increased with the distance. It was most marked in faint sounds. A contrary wind deadened the sound, but (and this was the most important result) all other winds deadened it too, though to a smaller degree. In still weather, or in a line perpendicular to the direction of the wind, the sound extended to its greatest distance. A commotion in the air is always injurious to the progress of sound, and this is intelligible if the gusts produce undulations in the air, which act on those of sound by the principle of interference. Derham made the same observation at Port Ferajo in Elba, *apropos* of the cannon of Leghorn, which was heard better in calm than in windy weather, even though the wind blew from Leghorn.

We may quote a remark of the Baron de Zach on this subject. This astronomer says that at the Seeberg Observatory, which is in a high and lonely situation, the sound of the neighbouring church bells, the noise of the mills, the barking of dogs, and the voices of men reached him clearly during the nights when the stars shone still and bright, while he could hear next to nothing when the stars trembled in the field of his telescope. Therefore the force of sound will indicate to a certain extent the state of the atmosphere.

The great difficulty in all these experiments is the want of an instrument to measure the intensity of sound ; one is obliged to trust entirely to the ear. Now the delicacy of hearing may vary day by day, and it is never the same in two different persons, and even the same person often hears better with one ear than the other, and, which is worse than all, the ear is more impressed by shrill tones than by deep.

One would have thought that the apparent intensity of a sound must be proportioned to the mechanical power employed to produce it, but it is not so. When a siren is turned by pressure of air from the bellows, the deep musical notes that it emits at first are far less piercing than the shrill notes produced as the velocity of rotation increases. The ear becomes more sensitive as the pitch of the note is raised, and it has been demonstrated that the high treble notes resound in the ear with a force beyond all others. Therefore it is certain that by the ear we can only compare sounds of the same quality. Should an exact measure for the intensity of sound be attempted, this is how it must be done:—The phonometer should be an instrument giving always an equal force to the sounds produced, by means of a constant pressure from a bellows. The distance must then be ascertained at which a sound from the phonometer would appear as powerful as that of which the intensity is to be tested. This intensity would be to that of the standard in the inverse proportion of the square of the distances of the phonometer and the sonorous source.

All movement which radiates freely—such as light, electricity, heat, and sound—spreads from its starting-point in concentric spheres. Thus, the surface of these spheres increasing always in proportion to the square of the radius, it follows that the intensity of force emanating from the centre must diminish in the same proportion as it is distributed over successive spheres. Hence the intensity of radiation decreases with the distance from the centre, in inverse ratio to the square of the distance. This law also governs gravitation; all the forces of attraction or repulsion submit to it. Theory would suggest that it should equally apply to sound. Delaroche and Dunal verified it in the following manner:—Having procured five bells, perfectly identical in tone, they placed one bell at one end of a straight line measured along the ground; the other four they hung at the opposite end. Standing midway between the bells, the sound emitted by the four ought to be four times as strong as the sound emitted by the one. Standing at one-third of the distance which separated the bells—that is to say, twice as far from the group as from the single bell—the observer found the sounds were equal. The law then was exact. The square of 2 being 4, and its inverse square $\frac{1}{4}$, the law requires that at a distance of 2 feet a sound should have only a fourth of the intensity which it possesses at the distance of 1 foot. Thus the sound of the four bells together being equal to 4, at the distance of 1 foot, ought to be no more than $\frac{1}{4}$ of 4, that is to say 1, at the distance of 2 feet. This the experiment proved, since at such a distance the four bells gave a sound equal to that of the single bell, at half the distance.

The distance at which the ear can distinguish sounds represents in some degree the measure of their intensity. The human voice is sometimes heard at a great distance. We have already told how in the Polar regions Foster was able to converse at a distance of 6,700 feet from his companion. Nicholson relates how, standing one night on Westminster Bridge, he heard the voices of workmen at Battersea, more than three miles off. The voices of the sentinels at Portsmouth may be heard at night in the Isle of Wight, five miles distant. The laughter of the sailors of an English man-of-war, stationed at Spithead, reached Portsmouth. It is hardly possible to credit Derham's affirmation that at Gibraltar the human voice has been heard above ten miles distant. Hinrichs assures us that a brass band may be distinguished at four miles, and the drum beating a retreat at Edinburgh Castle has been heard at nineteen miles. The report of a cannon travels very far, because it communicates a vibration to the soil. The cannonade of Florence was heard beyond Leghorn—that is to say, to a distance of about 56 miles—and that of Genoa to about 100. In 1762 the cannon of Mayence was heard at Timbeck, a small village about 148 miles off. In 1809 the booming of the cannon in Heligoland reached Hanover —157 miles; and on December 4th, 1832, the cannon of Antwerp was heard on the Erzgebirge mountains, 370 miles

Fig. 22.—Speaking-Trumpet.

distant. The eruption of St. Vincent in 1815 was heard at Demerara, 341 miles distant.

To increase the natural range of voice an instrument is often used, called in English a speaking-trumpet, in Latin *tuba stentorea*. It is formed of a conical tube, furnished with a monthpiece, and terminating in a wide-spreading cup; and is much used at sea to surmount the noise of wind and wave; and formerly the watchman used it to give warning of fires, or to call the labourers to their work in the fields.

It appears that the speaking-trumpet was invented by Samuel Morland in 1670. He had several models made in glass and copper, which were exhibited before King

Charles II. and Prince Rupert. In one experiment made at Deal, with an instrument about 5 feet in length and a diameter of 2 and 20 inches at its respective openings, the voice was carried over three miles.

When Morland's invention was made public, Father Athanasius Kircher claimed it on the pretext that he had

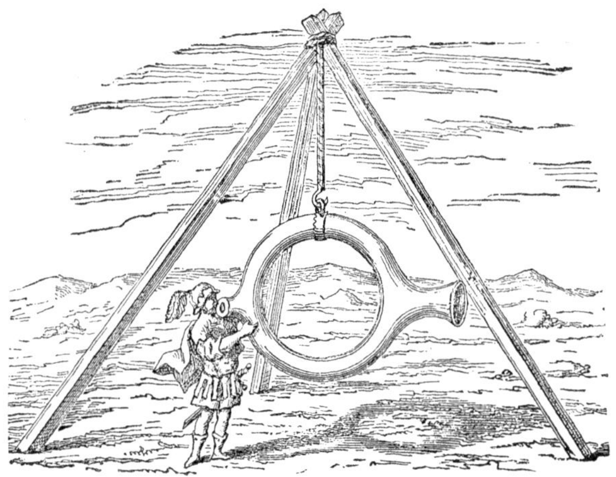

Fig. 23.—The Horn of Alexander.

already employed tubes of a conical form; but it is easy to see, from his earlier writings, that the learned Jesuit was speaking only of ear-trumpets. He gives in this connection a description of "The Horn of Alexander," from an old MS. entitled "Secreta Aristotelis ad Alexandrum Magnum," to be found in the Library of the Vatican. According to this unknown author, the horn enabled Alexander to call his soldiers from a distance of ten or twelve miles. The diameter of the ring must have been about eight feet. Father Kircher conjectures that it was mounted on three poles. Towards the end of the last century, a German, Professor Huth, wished to try the effect of such an instrument. He had a model constructed of thin iron plates, but on a somewhat smaller scale than that indicated by Kircher; and he found that a horn of this kind served as a powerful speaking-trumpet, especially when furnished with a widely spreading cup. In 1654, an Augustine monk named Salar made a similar trial, but no record was kept of the result.

Shortly after the invention of Morland, Cassegrain proposed that a hyperbolic form should be given to the speaking-trumpet. Conyers changed it to a paraboloid, and Jean Matthieu Hase made an elliptic mouth-piece, and a parabolic cup. All these plans (which have not stood the test of experience) suppose the increase of sound in the speaking trumpet is due to the interior reflection of the sonorous waves. This idea was enlarged on by Lambert in his theory of the speaking-trumpet, published in 1763, and quoted in almost all treatises on physics. It is laid down as a principle that the object of the instrument is to render the sonorous radiation parallel to the axis of the tube, wherefore the most suitable form must be chosen for realising this parallelism. Nothing could be more at variance with ascertained facts. According to the theory of reflections a cylindrical tube would be useless. Now, Hassenfratz has proved the contrary. The tick of a watch that in ordinary circumstances would be indistinguishable at a distance of about three feet, when placed at the end of a cylindrical tube twenty inches long, may be heard at a distance of six or seven feet. A cylindrical tube, furnished with an open cup or bell, would make a very good speaking-trumpet. Lambert thought the cup unnecessary. Experience proves the contrary: it contributes very sensibly to the increase of sound. Finally, Hassenfratz found that lining the interior of the trumpet with woollen stuff scarcely deadened the sound. Now, this lining must have prevented any reflection from the inner walls of the tube.

It results from these facts, that the augmentation of sound depends entirely on the geometrical form given to the column of air by the first impulsion. How is this influence exerted? No theory has yet explained the mystery. All that can be said is that the speaking-trumpet confines the sonorous waves, and keeps them from too soon dispersing, and as it were concentrates them. This notion makes us instinctively use our hands as a speaking-trumpet. The ancients used to fit a kind of cup or mouth-piece to the masks worn by their actors, to serve the same purpose.

Notice, still further, that sound is not augmented by a speaking-trumpet in the direction of its axis only. It is equally observable in every direction. Thus, if you speak through a trumpet at a certain distance from a high wall, the echo is almost equally powerful whether the mouth be turned towards the wall or in the opposite direction.

The tubes used on board ship are seldom more than six feet in length, and eleven inches in diameter. One was made in England of twenty feet or more, which carried words two miles. When used only for an inarticulate cry, a good speaking-trumpet will carry the sound three or four miles. Further experiments on this subject would be interesting.

In England and America they are trying many different

means for warning vessels at sea, when the lighthouses are invisible through fog. The common method is a bell. There is one on the Isle of Copeland, in the Irish Sea, rung by machinery, which may be heard at a distance of fourteen or fifteen miles. At Boulogne, a bell is fixed in the focus of a parabolic reflector, and struck alternately by three hammers, which are set in motion by a falling weight. On board some of the floating lighthouses they use drums or cannons. At New Brunswick they have a steam whistle. In a small island off Holyhead they protect the sea-gulls, that their cry may warn vessels; but, unfortunately, in 1856 the *Regulus* was wrecked in this part of St. George's Channel, and some rats escaping from the sinking vessel found their way to the island, and have multiplied to such a degree as seriously to affect the bird population. A cat was introduced to work havoc among the rats, but she made common cause with them, showing quite as great a partiality for the birds as they did.

The principal difficulty in this kind of signal is that the fog interferes with the propagation of sound—at least, it would seem so from Cunningham's experiments, but positive proof is wanting. To distinguish the signals of different stations they can employ intermittent sounds, or a succession of different notes. Cowper and Holmes have proposed steam trumpets for this purpose. Captain Ryder would unite a cannon with a whistle. It might be possible to propagate a very powerful sound through the water itself, in which case the sailors must use a long ear-trumpet, like that which Colladon had for his experiments on the lake of Geneva. They must fish for the sound. Prætorius invented an instrument of the same kind for the solid earth. It was a sort of shovel, driven into the ground; the ear, being applied to the handle, became conscious of a vibration at the approach of the enemy. The inconvenience of these contrivances is that they never tell the direction whence the noise proceeds.

When sound is propagated in a limited space of air, it loses but little in intensity. Of this, hearing-tubes afford a striking example. These are long tubes of metal or gutta-percha, by means of which conversation may be held between persons at some distance. They are much used in houses for communicating from the upper to the lower storeys, and on board ship for speaking to the man aloft, &c.

In the experiments made by Biot in the empty water-pipes of Paris, he found that the lowest sounds were perfectly transmitted through a column of air 3,120 feet in length. "Indeed," says he, "there was but one way to avoid hearing, and that was not to speak, even in the faintest whisper." The firing of a pistol at one end of the tube extinguished a lighted candle at the other, and blew some light bodies to a distance of twenty inches.

Once upon a time, in almost every fair might be found a "Delphic oracle," a Turk's head which answered all questions whispered into its ear. This was managed by a hearing-tube hidden in the pedestal of the apparatus, and communicating with a confederate. The most ingenious thing of the kind was M. de Kempelen's "speaking woman." This piece of wax-work was seated on a chair placed alternately in two different spots of the hall, where spectators were received. They spoke in her ear, and the answer seemed to come from her mouth. The plan of the thing was this: A tube passed from the hollow of the wax head through one of the feet of the chair. Two other tubes, connected with an adjoining chamber, passed under the flooring of the hall to two points, each marked by a small hole. Round these points the boards had been planed underneath to a very thin partition, and pierced with a small hole. They took care to place the chair so that the hollow foot covered one of these holes.

The "invisible woman," who created such a sensation at

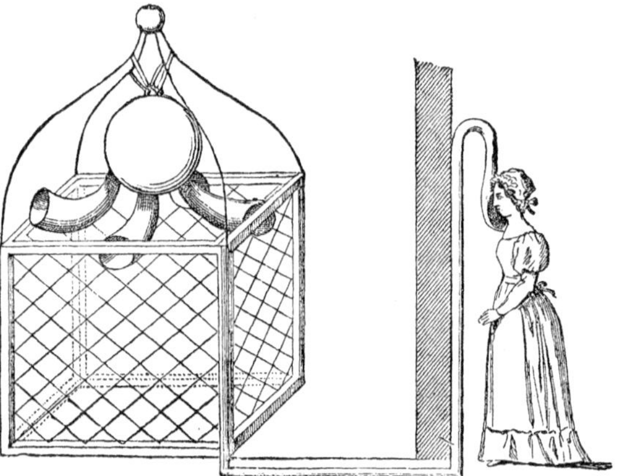

Fig. 24.—The Invisible Woman.

the beginning of this century in all the principal towns of the Continent, may be explained as simply. The most striking part of this machine was a hollow globe (Fig. 24), furnished with four horns in the shape of trumpets, and suspended by an iron bar, or more likely by four silk ribands, from the ceiling of the hall. This globe was enclosed in a cage of open trellis-work, sustained by four pillars, one of which was hollow. Through this passed a tube, which was carried also half-way through one of the upper horizontal cross-bars, whence the narrowest possible chink faced the opening of one of the trumpets. The voice seemed then to proceed from the globe. Probably the persons who gave the answers from a neighbouring room had a peep-hole by which they could watch what was passing in the hall. The questions were always put at one of the trumpets' mouths.

Sound is wonderfully propagated by means of chimneys, gas-pipes, heating apparatus, &c. Some chimneys will convey all kinds of noises from out-doors into the house; therefore in prisons and mad-houses they are specially careful, in the arrangement of such parts of the building, to avoid any possibility of communication between the prisoners or patients by such means.

At Carisbrook Castle, in the Isle of Wight, there is a well celebrated for its acoustic properties. When a pin is dropped down its contact with the water is distinctly heard, and shouting or coughing into the well produces a long-drawn echo. The depth of this well is 210 feet, and its diameter 12 feet.

In facts of this kind it is sometimes difficult to determine how much of the effect is due to the material of which the walls of the enclosed channel are formed. The same remark may apply to the transmission of sounds along a smooth surface. Hutton believed that a person reading aloud upon the Thames might be heard 118 feet off, while upon solid ground the distance must be limited to 75 feet. In the Argentine Theatre at Rome it has been noticed that the voices of the actors are much better heard since a water-pipe was carried under the flooring of the hall, and it is natural to suppose that the water has something to do with this improvement.

Most extraordinary acoustic effects may be noticed

under the domes of different churches, that can no more be explained by theories of the reflection of sound than the speaking-trumpets. The vaulted dome seems to guide the sound. It has been noticed that two persons talking in a whisper at opposite sides of a gallery under the cupola of St. Peter's at Rome can hear one another distinctly, without being audible to others. In the Whispering Gallery under the dome of St. Paul's, the same phenomenon occurs; the ticking of a watch even may be heard. In Gloucester Cathedral, a person speaking in low tones in the gallery east of the choir can be heard at the other end, 160 feet away. Brydone says the same thing happens in Girgenti Cathedral. When the great door is shut every syllable spoken near it reaches the other end of the nave, but cannot be heard midway.

These effects are but imperfectly explained by the reverberation of sonorous reflections, which accounts for the phenomena of elliptical vaults, as will be seen in the following chapter. It seems as though the surface guided the sound. Hutton tells how in a garden at Kingston a whisper along the wall was heard at a distance of 197 feet. It is still more striking to notice how a semi-cylindrical channel will guide sound. Hassenfratz put a watch at one end of a passage formed by two planks of wood resting edge to edge. He could then hear the beats at a distance of seventy-five feet, while in the open air they became inaudible at six feet. Some buildings have an accidental channel of this kind. There is one in an hexagonal hall of the Paris Observatory, where the opposite corners are furnished with a means of communication by a sort of gutter passing round the roof. A conversation may be carried on there in utter privacy, though the hall be filled with listeners. At the foot of the grand staircase in the Conservatory of Arts and Manufactures in Paris is a vaulted lobby, where the sound, following the arches, descends in the corners of the walls.

The same principle explains the mystery of "speaking chambers." Very often they are but the consequence of an accidental arrangement of the walls. We have the most curious of these phenomena in "The Ear of Dionysius," a cavern in the quarries of Syracuse, in Sicily. In the depths of this cave the tyrant of Syracuse had a cell formed for his prisoners, whence the least sound was carried to the ear of the sentinel watching at the entrance of the subterranean passage.

Here is Kircher's plan of the cave; *c* is the entrance, *d*

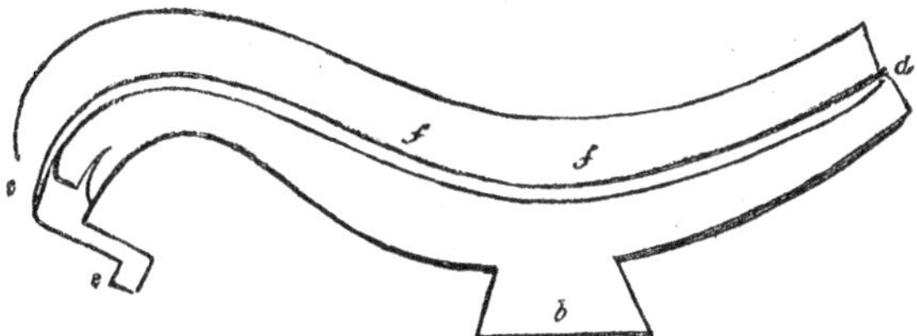

Fig. 25.—Plan of the Ear of Dionysius.

the cell; *f*, *f* is the projection of a large groove, thirty inches in diameter, hollowed in the middle of the roof, nearly 100 feet above the pavement, and ending at *e*, where the sentinel was stationed; *b* is a recess contrived in the side wall. The groove *f*, *f* acts as a sound-conductor. The opening *e* has long been walled up, and consequently at the present day the cave exhibits most curious effects of echo. Kircher visited it, and he tells how the faintest sound is exaggerated, so that a word pronounced in an undertone becomes a clamour, and a clap of the hands is like the report of a cannon. A duet sung by two voices is repeated as a quartet. The length of the cavern is about fifty-two feet.

Kircher has planned numberless imitations of "The Ear of Dionysius." Some consist of a large twisted tube, with a wide mouth opening towards the place where the sounds are produced, and passing into the interior of the room where the sounds are to be heard. This leads us to speak of the ear-trumpet, an instrument for gathering sound and condensing it in the ear. They are made in various forms, the simplest and the worst of which is the cone. It is requisite that the outer opening should be larger than the one introduced into the ear; then it is easy to understand how the movement in that portion of air which filled the wider mouth of the tube is concentrated in the narrower passage, and so reaches the ear intensified in power.

Towards the end of the seventeenth century, they tried ear-trumpets in the form of hunting-horns. One of the commonest forms is that given as 1 in the accompanying plate. No. 2 is another of the most usual. Curtis had some made to lengthen out like a telescope (No. 3). Ittard has devised numerous shapes. For instance, No. 4 is a kind of ellipsoid, furnished with a wide mouth, and a bent tube to fix in the ear. The dotted line shows a membrane of gold-beaters' skin, which renders the sound less confused, though it does not strengthen it. In No. 5 we have a shell, with a mouth and a tube added, and two membranes of gold-beaters' skin.

Quite recently Kœnig has constructed an ear-trumpet, which serves also as a stethoscope (No. 6). A capula, closed by a membrane, communicates with the ear through an india-rubber tube terminating in an ivory top. When any one speaks before this membrane it takes the impression from the motion of the air, and, carrying it to the air contained in the tube, forces it against the tympanum of the ear. When employed as a stethoscope, this simple membrane is replaced by a lens formed of a double membrane, that can be inflated by means of a cock at the side. The upper membrane is placed upon the chest of the invalid, where it moulds itself to the skin, and faithfully transmits the motion of the air imprisoned in the lens to the ear of the

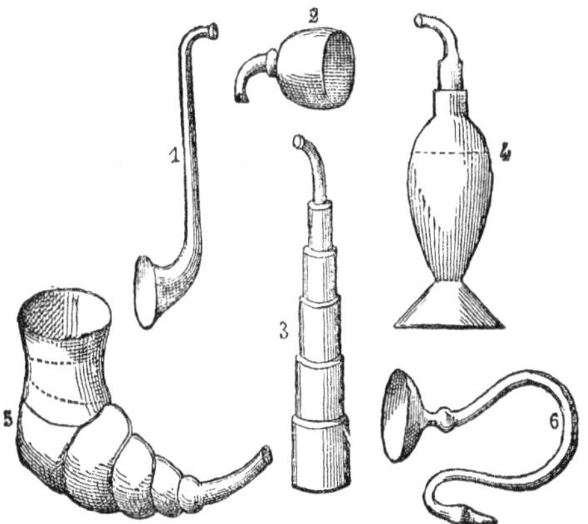

Fig. 26.—Ear-Trumpets.

doctor. With this apparatus the patient might sound his own lungs, by pressing the capula against his chest and introducing the tube into his ear; or a whole class of medical students might auscultate the same patient simultaneously, since several tubes can be introduced into the capula. The tubes may be lengthened to twelve or fifteen feet, without materially weakening the sound; so that a doctor could sit in his library, and listen to the beating of his patient's heart in an upper room.

CHAPTER V.

VELOCITY OF SOUND.

Mersenne—Bureau des Longitudes—Captain Parry—Regnault—Beudant—Colladon and Sturm—Biot—Wertheim—Distances by Sound—Depth of a Lake by the Echo from the Bottom.

THAT sound is not propagated instantaneously was noticed by the first inquirers into its phenomena. Every one knows that thunder is generally not heard till long after the lightning flash has passed, and the interval increases according to the distance of the storm. But what is the exact time that sound must take to travel a certain distance? in other words, what is its velocity? This was the question that Mersenne and Kircher set themselves to solve. "Light," says Mersenne, "spreads through the sphere of its activity in a moment; or, if it takes time, it is so short a time as to be imperceptible. But sound occupies time in travelling, which increases with the distance between the place of its production and the listener. This has been verified by many experiments. The axe of the wood-cutter will have struck a second blow before the sound of the first is heard at a distance of 600 paces. Repeated experiments are necessary to ascertain if this delay in sound is proportional to the distance." He then proceeds to describe the different experiments by which its velocity has been tested, such as counting the beats of the pulse from the moment when the flash of a musket or a piece of artillery is seen to the time when the report is heard. He records observations of this kind made at the siege of Rochelle by one of the officers; but the results are very inconsistent, and Mersenne therefore concludes that the velocity of sound varies according to local and atmospheric circumstances. Yet in any case he holds it certain that sound does not travel so fast as the ball from an arquebus; indeed he says, "The birds are often seen to fall from the branches of the trees before the report is heard, although one may be quite close to the arquebus."

In 1673 Kircher declared that nothing was yet known certainly as to the velocity of sound, but the Florentine Academy was instituting experiments for the purpose of throwing light on this interesting subject. These experiments seem to have taken place in 1660. They reckoned, from the time elapsing between the flash and the report of a cannon, that the velocity must be 1,175 feet a second. A simple means of gaining an approximate idea of the velocity of sound is found in an echo. Mersenne reckoned, with the help of a pendulum, that seven syllables could be pronounced in a second. Now, an echo at 519 feet distance will give back seven syllables. It takes one second to pronounce them, and they are heard again the following second. Therefore the sound travels 519 feet going, and the same returning—1,038 feet in all—in one second; "so that," says Mersenne, "we may take this as the velocity of reflected sounds, which I have found always the same, whether proceeding from trumpets, firearms, stones, or voices." These experiments, according to which the velocity of sound would be about 1,038 feet per second (we shall see presently that this number was pretty near the truth), were disputed by Kircher, who raised a host of objections against the supposed equality in time for the transmission of sounds of different kinds. He supposed that a very strong sound must of necessity be returned more quickly, just as a ball would rebound from a wall the faster according as the propelling force was stronger; but this comparison is altogether false, for sound does not rebound like the ball, since the mass of air in which the sound is propagated does not change its place. The air is not thrown against the obstacle, neither does it return to the ear; there is no analogy between the sonorous motion and that of the ball. Kircher also fancies that the echo is quicker in the silence of night than in the noisy day, and that the winds have something to do with the matter.

The first exact experiments on the velocity of sound in air were instituted in 1758 by a commission of the Academy of Science, composed of La Caille, Maraldi, and Cassini de Thury, who associated several others with them. They chose for their stations the Paris Observatory, the Pyramid of Montmartre, the Mill of Fontenay-aux-Roses, and the Château de Lay, at Montlhéry. Cannons placed on the heights of Montlhéry and Montmartre were fired alternately, and the observers at the four stations measured, by help of pendulums, the time which elapsed between the arrival of the flash and of the report. They found that on an average the sound took one minute twenty-four seconds to travel 93,140 feet—that is, about 1,106 feet per second, at a temperature of 6° (Cent.). Afterwards, when the influence of temperature came to be better known (augmenting the velocity about two feet for every degree Centigrade), they deduced from this reckoning a velocity of 1,093 feet for 0°. The observations made at the intermediate stations showed that the velocity of sound is uniform—that is to say, it does not slacken towards the end of its journey, however great the distance. They proved, moreover, that it is the same by day and by night, in fair weather and foul, and whatever may be the direction of the cannon-mouth; but that it is influenced by the wind, according to its force, and the angle that it makes with the direction of the sound. A contrary wind retards, while a favourable one accelerates its transmission. These experiments were repeated with some modifications by Kaestner, Benzenberg, Goldingham, and others, but their conclusions were not altogether satisfactory. A new measurement was taken in 1822, at the request of Laplace, by the members of the Bureau des Longitudes. Two pieces of cannon were placed, one on the elevation of Montlhéry, the other at Villejuif: the distance is about 61,067 feet. Prony, Arago, and Mathieu were stationed at Villejuif; Alexander Humboldt, Gay-Lussac, and Bouvard at Montlhéry. Each was provided with a good stop chronometer, recording at least the tenth of a second. The cannons fired at Villejuif were all heard at Montlhéry, but the return shots were so faint that few of them were heard at Villejuif. This singular circumstance prevented their noting the influence of the wind as accurately as they would have done. According to their calculation, the velocity of sound at a temperature of zero is 1,086 feet. For every degree of heat two feet must be added, so that at 15° the velocity would be 1,116 feet.

Since these memorable experiments, others have been made in Germany, Holland, America, and other places. During Franklin's voyage to the Arctic Seas in 1825, Lieut. Kendall discharged forty rounds of cannon, the temperature varying from 2° to 40° below zero. Captain Parry also made some observations on the propagation of sound in equally low temperatures. The united results of these inquiries tend to the conclusion that in calm air the velocity of sound is somewhere about 1,088 feet per second.

Biot contrived an ingenious plan for ascertaining whether sounds varying in pitch are equal in velocity. If not, it is clear that the notes of a musical air heard afar off would be altogether changed, since certain notes must be heard either too soon or too late. In ordinary circumstances this slight inequality would not be noticed, the distance of the instruments being insufficient to render such delay appreciable. Biot, therefore, arranged that a flute should be played at one end of the aqueduct of Arcueil (which was then empty) while he listened at the other. The melody reached him in perfect time and tune, having lost nothing in its transit through 3,120 feet of tubing.

About four years ago, M. Victor Regnault resumed these inquiries with all the appliances of modern science. Nearly 400 discharges of cannon were fired in the plain of Vincennes. The arrival of the sound was ascertained by means of a membrane, which, swinging a little pendulum at the arrival of the shock, thereby interrupted an electric current. The instant both of the flash and the report were registered on prepared paper by a Morse telegraph. On the same paper an electric pendulum marked the second, close by the spot where a vibrating tuning-fork registered the hundredth part of a second.

These experiments were terminated last year in the new sewers of St. Michel, a series of large tubes extending for a mile or more. The opening being closed the moment that the sound was thrown into the tube, it was observed that the noise of a pistol or trumpet rebounded, so to say, going backwards and forwards as many as ten times, swinging each time the pendulums placed along its route. M. Regnault also tried the effect of a simple shock or impetus given to a column of air without sound. I was present at some of these experiments, of which the results were curious enough; but as nothing has been yet published, it will be understood that I can say no more.

The velocity of sound in different gases has only been tested partially. It is believed that in oxygen, carbonic oxide, olefiant gas, nitrogen, and sulphuretted hydrogen it is about the same as in air; but in hydrogen, four times greater—that is to say, about 4,167 feet per second.

The velocity of sound in liquids was first experimented upon by Beudant. He had two vessels moored in the harbour of Marseilles, a certain distance apart. An assistant on board one of these vessels struck a bell sunk at its side, at the same time giving a signal which could be seen from the other boat, where the moment of arrival of the sound in the water was recorded. The velocity as determined on this occasion was 4,921 feet; but Beudant thought the result not worth publishing, on account of the imperfection of the means employed. It is only to be found in the Memoir of Colladon and Sturm. These two measured the velocity of sound in the water of the Lake of Geneva. The depth of the lake (459 feet), and the clearness of its waters, recommended it in a special manner for experiments of this kind. The greatest extent of deep water was found between Rolle and Thonon, a distance of about eight miles. A vessel was moored off Rolle, carrying a bell of nearly 140 pounds weight, which was submerged. It was so arranged that when the hammer struck the bell, a lighted match fell upon a heap of powder lying on the deck. Another vessel was moored off Thonon, from which they observed the flash, and noted the arrival of the sound by means of a curiously shaped hearing-trumpet

(Fig. 27). It was formed of a long tube, opened and bent, and had a membrane stretched across the mouth. The observer turned the surface of this membrane towards the bell, and placing his ear to the upper extremity of the cone, watched for the signal. The moment the flash was apparent he touched the spring of a "stop watch" (a kind of watch whose hands can be either stopped or set at liberty, by a simple pressure on the spring), stopping it immediately the sound reached him. This was invariably nine seconds after the flash. Dividing the distance between the two vessels by the interval of time, the velocity is determined to be 4,708 feet per second—more than four times greater than in the air.

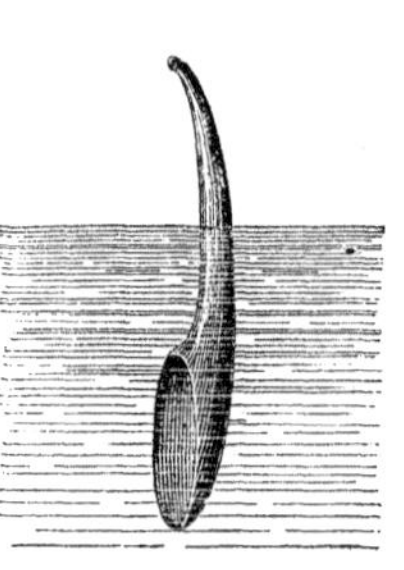

Fig. 27.

These experiments gave rise to many interesting remarks upon the propagation of sound in water. Instead of the prolonged resonance that is produced in air, the sound of the bell was short and flat, like the clashing of two steel blades. The water, which is but slightly compressible, had robbed it completely of its ringing tone. At one time the lake was rough and stormy, and they had great trouble in keeping the boats to their moorings, but this had not the slightest influence on their experiment. Wertheim afterwards determined the velocity of sounds in different liquids, and of his results, these are the two extremes: in absolute alcohol, at a temperature of 23° C., the velocity is 3,808 feet per second; and in a solution of chloride of calcium, 6,496 feet.

Through solids the transmission of sound is much more rapid than in gases or liquids. The early experimenters who tried to measure its velocity by laths of wood, cords, &c., found it too great to be appreciable. The efforts of Hassenfratz were fruitless. Biot and Martin tried by means of the iron pipes made to carry the waters of the Seine from Marly to the aqueduct of Luciennes, and found that from a small bell hung at one extremity, two sounds reached the ear in succession. The first was transmitted through the iron, and after an interval of between two and three seconds was followed by another through the air. From this experiment they calculated a velocity of 8,859 feet. This deduction is not correct: the result is too small. This may be explained by the lead used at the joining of the pipes interrupting slightly the transmission of the sound. When Breguet and Wertheim afterwards experimented on the telegraph wires of the Versailles Railway, they gave as the result of their calculation 11,434 feet per second for the velocity of sound in iron wire. By the method of vibrations (an indirect mode) Wertheim determined the velocity of sound in some metals. In lead it is equal to four times what it is in air, about 4,265 feet; in silver and platinum, 8,859 feet; in zinc and copper, 12,140 feet; in iron and steel, 16,404 feet. The highest known velocity of sound is that which Chladni found in the wood of the fir-tree, about 19,685 feet—eighteen times that of transmission through the air.

To form a comparison of these different results, let us imagine for a moment that the stone tunnel projected by M. T. Gamond is constructed under the English Channel. The distance from Cape Grisnez to Eastware Point (the proposed stations) is about twenty-one miles. A cannon fired at Grisnez would be heard at the English station in ninety-seven seconds, through the air; the sea-water would transmit the sound in twenty-three seconds; by the iron rails it would come in a little over six seconds, and by the telegraphic wires probably a little faster. Finally, if there were a lath of fir-wood long enough to join the opposite shores it would transmit the sound in five seconds.

The velocity of sound in air being known, we can employ it as a means of approximately measuring distances. Every second that elapses between the flash and the report of a firearm represents a distance of 1,116 feet between the station where it is fired and the position of the observer. We have already seen how Mersenne made use of this fact. M. d'Abbadie measured different sites in Ethiopia by the same means. In the island of Mocawa, during the Ramadam (a religious fast of the Mussulmans), a cannon was always fired at sunset announcing the end of the day's fast. M. Antoine d'Abbadie took the opportunity of noting the time which passed between the flash and the arrival of the sound on the opposite bank. He took his station on a hill near the village of Omkullu, and there awaited the report of the cannon. The sound reached him eighteen seconds after he saw the flash: the distance he reckoned to be 21,129 feet. Another time he measured in the same way the distance from the town of Aoua to Mount Saloda. His brother Arnauld took his stand on the mountain, while he himself was upon the roof of a house in the town, armed with a blunderbuss. They fired alternately, and each one marked the seconds by his watch: the distance was found to be nearly two miles. But it seems the brothers made too much noise, for they were both banished from the Tigris.

Newton gives a formula by which to calculate the depth of a well from the time that passes between the moment when a stone leaves the margin and that at which it is heard to strike the water. Ten seconds would give a depth of about

1,247 feet. We might get the depth of a lake, or even of the sea, if we could note the reflection of a sound strong enough to be returned from the bottom. Arago proposed this to Colladon in 1826, but it was never tried till 1838, when at the request of the Admiralty of the United States Mr. Bonnycastle made the experiment. The American professor found that sound was better perceived in the water than in the air, and that the greatest distance at which a bell could be heard under the water was two miles. These conclusions were disputed by Colladon, who urged his experiments made on the Lake of Geneva. In 1826 he had heard a bell of nearly 140 pounds weight at a distance of eight miles. In 1841 a bell of about 180 pounds, lent by one of the churches of the canton of Geneva, was heard at a distance of twenty-one miles. It was suspended forty-nine feet under water, and the hammer which struck it weighed over twenty pounds; from which Colladon concluded that, under favourable conditions, sound would be propagated under water to a vast distance. The noise of the paddle-wheels of a steamboat is not heard beyond 3,000 to 4,000 feet under water; but the noise of a chain shaken at a certain depth is so distinct, that a ship at two miles distance may be heard to weigh anchor.

It is understood that in these experiments it is always necessary to use a hydro-acoustic trumpet. During the trial of the great bell, each blow could be heard in a house built upon an embankment at a distance of two miles, although the house was separated from the bell by a promontory: the sound seemed to come from the foundations and the walls. Colladon says nothing of the possibility of measuring the depth of water by an echo from the bottom.

CHAPTER VI.

REFLECTION OF SOUND.

Laws of Reflection—Echo—Polysyllabic Echo—Polyphonic Echo—Heterophonic Echo — Reflection and Resonance — Celebrated Echoes—Legends—Refraction of Sound.

THE laws of reflection show a perfect analogy between light and sound. Sounds are reflected, like luminous rays, from any obstacle they may encounter; and just as we find the polished surface of a mirror giving back more light than a rough surface, so different substances return the sonorous waves with more or less force. Hard and solid bodies reflect much better than soft and flexible ones, that cannot easily right themselves after pressure.

The laws of the reflection of sound do not appear quite so simple as those which govern the movement of luminous rays, for the sound-waves travel in curved lines, bending round obstacles. Nevertheless, for the sake of simplifying our explanation, we may be permitted to speak of sonorous "rays" just as we speak of luminous rays, meaning thereby the direction in which a sound arrives in greatest power, when propagated through the air. Therefore we may say for sound as for light, that the incidental and the reflected ray make equal angles with the reflecting surface, and that they are comprised in a plane perpendicular to this surface. The same law obtains in the shock of elastic bodies. Billiard-players know that the ball rebounds from the cushion in a direction symmetrical with that in which the propelling force was given. It is thus that a voice striking the wall M, in the direction A, M, is thrown back in the direction M, B, symmetrical with the first as regards its relation to the wall, or (which comes to the same thing) to the perpendicular M, N. The angle which this perpendicular makes

Fig. 28.—Reflection of Sound.

with A, M, is called the angle of incidence; that which it makes with M, B, is the angle of reflection. These two angles are always equal, and the reflected ray M, B, is always in the same plane as A, M, and the perpendicular M, N.

When the point A, whence the sound emanates, approaches the line M, N, the point B, towards which the sound is reflected, approaches it also, and these two points coincide when the sound travels in the direction of the perpendicular. That is to say, a voice thrown out at N, and striking the wall in the direction of the perpendicular N, M, would return by the same path from M to N.

These principles will help us to understand the phenomena of echoes, as we call the repetition of a sound when reflected by some distant object. Let us suppose, to begin with, that we have but one reflecting surface. If the observer wishes to hear the echo of his own voice, he must place himself on the line M, N, which is perpendicular to the reflecting surface. If he wishes to hear the echo of a noise produced at a point A, he must place himself at a point B, symmetrical with reference to the perpendicular M, N. Before hearing the reflected sound, which travels by the broken line A, M, B, he will of necessity hear the sound which passes direct from A to B, since this has a shorter road to travel. We assume, of course, that no obstacle stands between these two points which could impede its passage. The observer will then, in general, hear two sounds in succession, if the first has ceased before the arrival of the second. This is a necessary condition for a distant echo, and it evidently depends on the distance of the reflecting surface.

We will first consider a case where the sound returns to the place of its departure. The observer is then at N; he hears his own voice first at the moment of utterance, then again after the sound has travelled twice the distance M, N. Now, it takes at least one-tenth of a second to pronounce one syllable, and that is pretty quick speaking; on an average we do not pronounce more than five syllables in a second. If, then, the reflecting obstacle be too near the observer, the first syllable will return before he has uttered the last, and there will be confusion, the last syllables only, or perhaps none at all, being given distinctly.

We have seen that sound travels on an average about 1,116 feet per second; in the tenth of a second, then, it would be 112 feet nearly, and 224 feet in the fifth of a second. An obstacle distant 112 feet in a straight line would, therefore, send back the sound after one-fifth of a second, allowing one-tenth for its journey there and one-tenth for its return. This distance would suffice for a monosyllabic echo—that is to say, for the repetition of a single syllable. One-fifth of a second it would take to pronounce it, so that as I pronounce the end of my syllable the beginning has already returned to me. If the obstacle is nearer than 112 feet, the reflected sound breaks in upon the articulated sound and confuses it. If the obstacle is at a greater distance, a longer or a shorter time will elapse between the spoken syllable and the echo which repeats it.

All that has been said of the monosyllabic echo will apply to the polysyllabic, or the echoes of many syllables. We have only to increase the distance in proportion to the number of syllables to be repeated. For two syllables we must allow 224 feet; for three, 336 feet, and so on. Of course, if more than five syllables are uttered in a second, the distance allowed may be smaller; but if less than five, the distance must be greater. The principle is always the same. The distance must allow the sound to go and return during the time taken for the utterance of the phrase. However, it is true that several syllables spoken in succession are produced more rapidly than a single one is apt to be, which explains why Kircher found the distances decrease slightly for polysyllabic echoes. Whilst he gave 100 feet for an isolated syllable, he reckoned only 190 for two, and 600 for the seven syllables—

"Arma virumque cano."

He states elsewhere that the distances allow of a great latitude. The echo of a trumpet is distinct from 90 to 110 feet, and the distance for an echo of seven syllables may be reduced to 400 feet, while sometimes 600 feet is not sufficient for the repetition of seven syllables. When too many syllables are given for the echo to repeat distinctly, the first which return are drowned by the last uttered, and only a mutilated edition of the phrase is obtained. From this circumstance it is easy to hold a conversation with the echo, by question and answer, remembering only that the end of the question must serve as reply.

Cardan tells a story of a man who, wishing to cross a river, could not find the ford. In his disappointment he heaved a sigh. "Oh!" replied the echo. He thought himself no longer alone, and began the following dialogue:—

"Onde devo passar?"

"Passa."

"Qui?"

"Qui."

However, seeing he had a dangerous whirlpool to pass, he asked again—

"Devo passar qui?"

"Passa qui."

The man was frightened, thinking himself the sport of some mocking demon, and returned home without daring to cross the water. He told his adventure to Cardan, who had little trouble in explaining it.

We have supposed hitherto that the observer heard the echo of a sound produced by himself, returning by reflection to its point of departure. The same reasoning applies in cases where the sound rises at a certain distance from the observer, as in Fig. 28, where the listener is placed at B and the sound comes from A. We only have to consider the difference of the direct road A, B, and of the indirect A, M, B. This difference represents the circuitous route by which the reflected sound has travelled, or the advance gained by that sound which has travelled direct; and it must be equal to twice 112 feet—that is, to 224 feet—for a monosyllabic echo; double that for an echo of two syllables, and so on.

Lastly, there are multiplied or polyphonic echoes—those

Fig. 29.—The Heptaphonic Echo.

which reproduce several times consecutively the same sound or phrase. They are caused when several obstacles placed at different distances, acting either alone or together, send back the sound in successive echoes. The accompanying figure represents a heptaphonic echo, that is, of seven voices. The projecting pieces of the wall, A, B, C, D, E, F, G, which throw back the sound, are at *nearly* equal distances; it returns first from A, then from the others in succession. If the echo have to repeat a single syllable seven times, the successive distances must differ by at least 112 feet; for two syllables,

224 feet, and so on. The farther the distance, the feebler the echo, as the sound is scattered and lost; thus the voice dies gradually until it completely ceases. When the obstacles which produce the echo, instead of being placed at equal distances, are nearer together in proportion as they recede from the observer, the echoes mingle, the second arriving before the end of the first, the third before the second is

Fig. 30.

completed, and so on. Kircher shows how this law may be used to produce a sentence from a word. Suppose a five-voiced echo so disposed (Fig. 30) that the first repeats distinctly the word *clamore* (voice): if the second obstacle be at double the distance, the third at triple, and so on, a trisyllable in five sounds would be produced. But put the second so near that the sound of the consonants *c l* is lost in the first echo, only the word *amore* would be heard; and by placing the succeeding obstacles at properly arranged intervals, the third would repeat *more*, the fourth *ore*, the last *re*. So that, in asking in a loud voice the question, *Tibi vero gratias agam, [illegible] o clamore?* the echo replies, *Clamore—amore—more—ore—re.* The word *constabis* would divide into *stabis—abis—bis—is*, but without giving a sentence of any meaning.

Fig. 31.—The Heterophonic Echo.

Kircher also tried to construct an heterophonic echo—one that should reply in a different word—which he contrived thus: At the salient angle of a wall was an obstacle (Fig. 31) which, instead of throwing the voice back to the spot from whence it had come, sent it round to the other side of the building, where an accomplice was concealed; he replying, his voice followed the same route as the question, and reached the mystified hearer, who, having heard asked, *Quod tibi nomen?* (What is your name?) hears the answer, *Constantinus.* Kircher relates that he had been much amused at his friends' expense by this innocent mystification, in the Campagna at Rome. To render the illusion complete, the voices of the two actors should be alike.

It would be possible to utilise the echoes of a church, as ornaments to the singing, by disposing pauses which should be filled in by their resoundings. Kircher gives several examples of musical effects thus obtained, and adds that the churches of St. Peter and St. James of the Incurables at Rome are particularly adapted for the application of this artifice.

The Hebrew name for echo is "daughter of the voice;" to the ancient poets Echo was a nymph who loved the beautiful Narcissus, whose love being despised, she dissolved in tears, and remained only a voice which replied to the passion of another—

> "Nec prior ipsa loqui didicit resonabilis Echo."

The echoes which animate a landscape seem to establish a kind of sympathy between man and nature. The forest partakes in our joys, and repeats the cries of the hunters and the notes of the horn.

> "Non canimus surdis, respondent omnia silvæ."—VIRGIL.

As Mersenne says, God has given a voice to the woods, rivers, and mountains.

The echoes in towns, and regions of peculiar conformation, are of various qualities: sometimes the response is muffled and hoarse, sometimes clear and distinctly accented.

These differences in quality, depending evidently on the character of the reflecting surfaces, prove that an echo is something more than mere reflection. It is beyond doubt that the phenomena of resonance, of which we shall speak subsequently, play a certain part in it. All the facts observed prove, also, that the reflection of sound can be made clear and distinct from very irregular surfaces; an old rampart, a ruined tower, a tree, a hill, a wooded gorge, are the obstacles which form the best echoes. The luminous image is perfect in proportion as the surface which reflects it is uniform; the sonorous image is not subject to these conditions. We must conclude that in most cases the mode of action of the surfaces which form an echo has some analogy with the effects of curved mirrors. Perhaps the resonance of the obstacles themselves, and of the air confined in them, contributes largely to the production of the phenomenon.

It is certain that the concurring conditions which should be regarded as favourable or necessary to the production of an echo, are far from being known. Theory and experiment are equally at fault. In some cases the local conditions which should, according to the theory of reflections, produce an echo, do really produce it; but often our expectation is deceived where no reason for it can be discovered.

The echoes of forests depend much, probably, on the grouping of the trees, as the following facts may show:—

Gay Vernon, in his youth, had often amused himself by waking an echo formed by the buildings of a mill. After passing several years in Paris, he returned to his native village: to his surprise the echo no longer existed; yet nothing had been changed about the mill—only a group of trees, which formerly shaded it, had been cut down.

In the plain of Montrouge, near Paris, there was formerly a remarkable echo produced by a wall, before which were several rows of trees. Hassenfratz tried to ascertain on what circumstances the phenomenon depended. He placed an assistant at a certain distance to call out, and approached the wall slowly, listening carefully: the echo died away as he drew near, but there remained a faint sound, proceeding not from the wall, but from the trees. Putting his ear to their trunks he perceived a slight tremor, while in the wall there was no vibration. He also observed that the walls of certain houses produced an echo when the windows were shut; or with the windows open, but the doors shut. In some vaults certain notes only produce the effect. The echo of the ancient college of Harcourt has a strange peculiarity: it returns the voice of a man placed in the middle of the court, but the low notes were heard in the direction of the Rue de la Harpe, the high notes in a direction fifty degrees more to the north.

All these facts show that Echo is a capricious being, whose caprices are not easily divined. Here is a story in illustration: — An Englishman, travelling in Italy, met with an echo so beautiful that he determined to buy it. It was produced by a detached house. This was taken down, carried to England, and reconstructed on one of his estates, exactly on its original plan—a place having been chosen for it at exactly the same distance from his dwelling as it stood, in Italy, from the place whence the echo was most distinctly heard. To test the echo he sent for a box of pistols, charged both the weapons, went to the window, and fired—no sound was returned; drawing the trigger of the second, he shot himself through the brain! It was never known what defect in the construction was the cause of this lamentable disappointment.

Clouds also re-echo terrestrial noises. The members of the Bureau des longitudes, in the course of their experiments for measuring the velocity of sound, found that the report of cannon was always followed by an echo if clouds were overhead. The rumbling of thunder is owing partly to the multiplied reflection of sound between the earth and the storm clouds. Echoes are also produced by excessively high waves, and the sails of ships, and it is said that words spoken through a speaking-trumpet come back if they strike on the convex surface of the sails.

Echoes are especially distinct in the silence of night; the noises of day prevent their being heard distinctly. Mersenne relates that the echo of Ormesson, in the valley of Montmorency, replies fourteen syllables at night, and only seven in the day-time.

In deep valleys, and the hollowed strands of rivers, remarkable echoes are found. In one well-known echo, between Coblenz and Bingen, where the waters of the Nahe flow into the Rhine, there is an echo which gives seventeen repeats, the voice seeming alternately far and near. One day, the steamer not having the usual fire-arms on board to rouse the echo for the amusement of the tourists, there were loud cries for a pistol. A Pole, not understanding the case, rushed on to the bridge, exclaiming, "I have no pistol, but here is a dagger." Ebell relates that an echo at Derenberg, near Holberstadt, repeats the twenty-seven syllables of this sentence — *Conturbabantur Constantinopolitani innumerabilibus sollicitudinibus.* It would be as astonishing to find a mouth capable of pronouncing them quickly; but as he says that the distance was only 254 paces, which is not enough for such an echo, there must be some mistake in the account.

It is said that near Brussels there is an echo of fifteen repeats; and at Rosneath, near Glasgow, on the Clyde, one which repeats an air of music three times. This scarcely seems credible.

An echo at Woodstock, near Oxford, repeats seventeen times by day, and twenty times by night; the distance is half a mile.

At Genetay, two leagues from Rouen, in a semi-circular court, there is a remarkable echo. When crossing the court singing, the singer hears only his own voice, while those listening hear only the echo, single or multiplied, according to their position.

At three leagues from Verdun are two towers, apart, and isolated from the building to which they belong; standing midway between them, the speaker's voice is echoed twelve or thirteen times with decreasing force, but except from this spot the echo is lost; while between one tower and the building a single echo is heard. Near Heidelberg is an echo which imitates thunder. To waken it a pistol is fired from the base of the hill Heiligenberg; a wooded gorge in front so reflects the sound, that instead of the report of the pistol a noise of thunder is heard.

In Bohemia, near Aderbach, there is a circular space six leagues in diameter, surrounded by bare pointed rocks. At one spot in the centre is an echo which repeats three times a sentence of seven syllables, while at a short distance off no echo is perceived.

In the walls of Avignon, Kircher found the voice repeated eight times. In Rome an echo is repeated from two to seven times. Boissard, in the "Roman Topography," gives this description of the tomb of Cœcilia Metella: It is a round tower, its walls twenty-four feet

Fig. 32.—HALL AT SIMONETTA, NEAR MILAN.

thick, and ornamented with 200 heads of bulls in marble, to commemorate the two hecatombs sacrificed at the funeral of the daughter of Metellus Crassus. This monument is situated near St. Sebastian, and called "The Bull's Head." A sentence spoken at the base of the hill on which it stands produces a multiplied echo. Boissard says that when he sang the first line of the Æneid, it was repeated eight times distinctly, and several times more imperfectly. Mersenne, speaking of this echo, says the place can still be seen in which the hecatomb was immolated, where the echo would make the sacrifice seem larger than it was. Whether the place was chosen to give a greater solemnity to the rite, or whether it was chosen for the burial-place of the house of Crassus to immortalise it by multiplying their names to posterity, he could not tell.

In a private dwelling an echo is not at all pleasant, as it causes what is said or done to be heard at a distance. It is only in large halls or places of amusement that it would be desirable, while in a church, if it makes the preacher's voice better heard, it also frequently interrupts him by the re-echo.

The drawing by Kircher (Fig. 32) represents the Hall at Simonetta, near Milan. Measured from the interior of the court, the façade is 121 feet, and the wings 66 feet; the height of the upper storey, between the gallery and the roof, 32 feet, the gallery 16 feet. When a pistol is fired from the window in the left wing, it is repeated forty or fifty times, and the sound of the voice twenty-four to thirty times. Addison and Monge tried it, and Bernouilli believed he counted sixty repetitions.

In vaulted buildings there is an echo, owing its peculiarity to the laws of geometric curves. The ellipse is a lengthened curve like a flattened circle, and two points in its interior, *f*, *f* (Fig. 33), are called the foci, because in each of them are collected the rays of light or sound, which, diverging

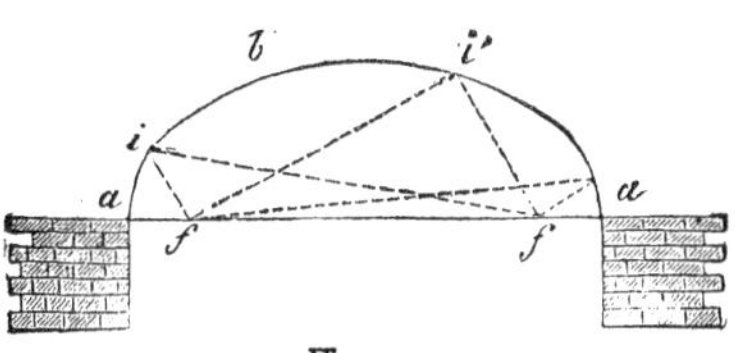

Fig. 33.

from the other, are reflected from the interior of the curve. A person placed at one of the foci of an elliptic curve hears a whisper from the other focus, so that two persons placed at

Fig. 34.

these positions could converse in a whisper without being overheard. There is a building of this kind at Muiden, near Amsterdam. Parabolic surfaces have one focus, to which parallel rays converge after reflection, while those diverging from it become parallel after reflection; so that if two parabolic mirrors are placed opposite each other, the slightest sound made at the focus of one is heard at the focus of the other, as is shown in Fig. 34. This makes them applicable in lighthouses, for throwing rays of light or the sounds of bells to a distance. With less reason they are chosen for acoustic trumpets. It is supposed that at the focus where the ear is placed the rays coming from a certain distance are condensed, as a parabolic mirror condenses at its focus the sun's

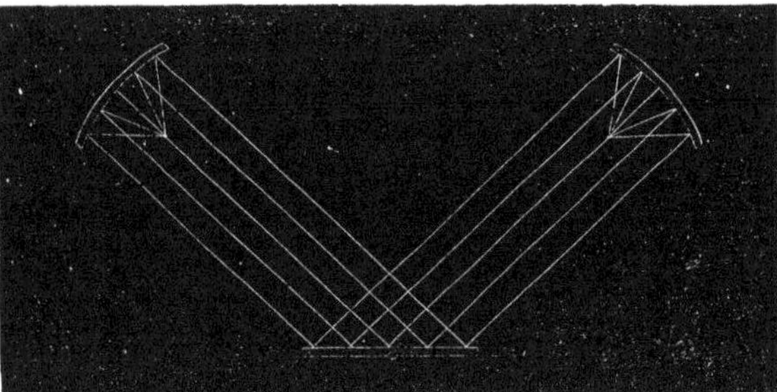

Fig. 35.

rays. The sails of a ship sometimes produce this effect when inflated by the wind. Arnott says that in a coasting vessel off Brazil, by standing before the mainsail the bells of San Salvador could be heard from a distance of 110 miles.

Church vaults, caves, and ramparts very often furnish some curious illustrations of acoustic effects. In an elliptic vault, sound issuing from one point is heard at another fixed point by a single reflection from the wall; and between two opposite parabolic vaults it is heard, though less distinctly, by means of a double reflection. Other systems of curves might give the same result by a number of successive reflections.

Thus two parabolas combined with a plane surface, as in Fig. 35, would give it by means of a triple reflection; and it is possible that the action of multiplied reflections would go far to explain many curious results.

Sound is much increased by the echoes in a closed vault. In a cave of the Pantheon, the keeper by striking the flap of his greatcoat makes a noise like the report of a cannon. The same phenomenon is found in the caves of Kentucky. In the Cave of Smellin, near Viborg, in Finland, by throwing in a live animal you hear terrible noises. Olaus Magnus says that when an enemy approached the inhabitants would conceal themselves, while the boldest amongst them cast an animal into the cavern, whose terrible roarings "overthrew the enemies like oxen at the shambles, when the Finlanders leaving their hiding-places spoiled the slain." Pliny tells of a similar cave in Dalmatia, where the falling of a stone raised a perfect storm. Fingal's Cave, in the island of Staffa, presents another remarkable phenomenon. The end of this cavern is dark and gloomy, and may be compared to the chancel of a church, while the basaltic columns may be likened to the organ-pipes. At the extremity of the grotto, and near the level of the water, is a small opening, whence come harmonious sounds, which are produced by the swell rising and falling.

St. Clement of Alexandria relates that in Persia were three mountains in an open country, so situated that approaching the first you heard confused voices and wrangling; on nearing the second the hubbub increased, but reaching the third you heard sounds of mirth and rejoicing.

The panic terror which overcame the Gauls near the temple of Delphi, defended by the god Pan, is attributed to echoes. In the same way, Mersenne says, "The Persians, while ravaging Greece and Megara, awaking an echo in the night, imagined they heard the cries of numerous enemies, and attacked the resounding rock, on which they spent their courage and their darts, and were next day taken captive."

Another remarkable analogy between light and sound is the refraction which both rays undergo in passing from one medium to another. A spoon put into a glass of water seems to bend; this is the effect of refraction. The rays of light which meet the water in an inclined direction are bent when they emerge into the air. The effect of prisms and lenses depends on the successive refractions to which light is subject on passing from the air into the glass, and back again; the glass being so prepared as to give the requisite deviation. M. Hajech thus proves that the rays of sound follow the same laws: He had a hole made in the partition wall of two rooms, and placed in it a tube closed by two membranes. This tube was successively filled with water, carbonic acid, hydrogen, ammoniacal gas, &c. At one end another tube was attached, filled with air, and ending in a wadded box containing an alarum watch. The sound passed through the tube containing the gas or liquid, and the observer in the next room noticed where the sound had most force. When the two membranes were perpendicular to the axis of the tube, the direction corresponded with the axis, without deviation; but when the front membrane was inclined to the axis, a sensible deviation was perceived, which was measured by holding a plumb-line to the ear, which traced the arc of a circle on the floor. These experiments showed that the rays of sound are refracted by the same laws as the rays of light: they depend on the angle at which the rays strike the reflecting body, and the comparative velocity with which they are transmitted through the two media.

This was the same for water and hydrogen, but different in the case of carbonic acid.

M. Sondhauss observed the refraction of sound by means of a lens of collodion filled with carbonic acid. When a watch was placed in the axis of this lens, the sound was concentrated at another point of the axis on the opposite side. This, therefore, was the focus, and the sound of the watch was distinctly heard; but when the lens was removed it was lost. This experiment was made more easily by means of Helmholtz's sonorous globe; this was moved slowly before the lens, and the india-rubber tube attached to it placed in the ear.

Mersenne has also considered the question "whether sounds are bent by refraction, as light is when it passes from one medium into another." But he only explains how light is affected by refraction, and hence how magnifying lenses should be cut; and then adds, "I do not believe that these effects can be produced in sounds by human industry; as to the angels, if they like to dispose the vibrations of the air as they please, I do not doubt they can do the same thing with sound as with light."

CHAPTER VII.

RESONANCE.

Resonance—Vitruvian Vases—Harmonic Tablets—Sonorous Globes —Glasses Broken by the Voice—Acoustics of Churches and Theatres.

THE assertion that sound passes round an obstacle must not be taken too literally. Very massive bodies may arrest it, as an opaque screen does light. Two persons separated by a rising ground can hear each other, because sound passes over the obstacle as light cannot; but it is made fainter, and they would hear much better if it were removed. It is only when sound is conducted through a closed tube or passage that its direction may be changed without diminishing its force; in open air it grows fainter, as daily experience proves. To hear a speaker well you should be in front of him, and to hear an indistinct voice you instinctively turn the ear in the direction from which it proceeds. When the stream of sound meets an obstacle it can turn the other way, as a current checked by an island, but its force is diminished.

A very large and massive object will entirely arrest sound. Under the arches of large bridges you may place yourself so that no sound from without can reach you. Behind the vertical fall of the Rhine at Schaffhausen there is complete silence. The sounds of bells may often be heard in streets from an opposite direction to that of the bells; the houses arrest their sound, and only the reflected sound from the opposite walls is audible.

Elastic bodies of slight density offer little impediment to the passage of sound, and are of little use in damping it. It would be as wise to try to keep out light with a glass screen as to try and intercept sound by a boarded partition. The elastic body becomes itself sonorous and vibrates to the touch. The same thing is observed when sound is reflected from an elastic surface, which acts as a spring-board to return the sound with vigour. By this means the wonderful intensity of some echoes is to be explained. At the same time other sounds, arising themselves from the reflecting surface, mingle faintly with the reflected sound. We say then that the surface resounds. It is analogous to the reflection of the solar rays by a body which, besides returning the direct rays, becomes heated and then radiates heat in all directions.

The *resonance of arches* is a complex phenomenon due both to resonance and reflection. Sound returns too quickly from the walls of a high arch to produce a distinct echo, yet not quickly enough to be blended with the original sound; the vibrations of the walls bring in a new element, and so a thousand confused noises are produced, which give rise to the remarkable effects we have noticed in speaking of echoes. We can observe these phenomena when passing in a steamboat under a bridge, of which the sides and arches intensify the sound of the paddles. When a locomotive rushes with great velocity underneath a bridge, the reflection of the noise produces a sound like violent explosion, and in a tunnel the uproar becomes deafening. Sheets of water are very favourable to these effects, by the facility with which they reflect sound. Thus Cagniard de Latour, having compared two pits—one dry, the other containing a little water—found the latter was much more sonorous than the former. Under the arches of bridges, the resonance is sensibly weaker when there is no water underneath.

Drapery, tapestry, and all fabrics of that class have the effect of deadening sound; they destroy sound in a room which contains them just as gloomy colours render it dark. It is for this reason that even a good piano is often not well heard in a room carpeted, hung with curtains, and filled with cushioned furniture. Empty rooms are always remarkably sonorous. In churches, or in rooms where meetings are held, too great resonance is very injurious to distinct hearing. It is apt to drown the voice of the speaker and render him unintelligible. But the case is very different in a concert-hall; there we endeavour to increase the resonance of the walls by a casing of thin wood.

In the time of Rousseau the best constructed orchestras were to be found, it is said, in the Italian theatres. The platform was made of light and resonant wood, such as pine. It was supported upon arches with an empty space beneath, and it was separated from the audience by a railing in the pit distant a foot or two from it. By this arrangement even the body of the orchestra was supported freely and could vibrate without obstacle, thus allowing full scope to the power of the instruments. At the Paris opera, on the other hand, the orchestra was very badly arranged, being near the ground and enclosed all around with massive wood and iron, which destroyed all resonance. At the present time, the principles of construction so much praised by Rousseau are adopted in the majority of theatres specially devoted to music; but it is true that many competent architects consider them useless or even injurious.

Vitruvius tells us that the Greeks put inverted brass bells over conical supports in the niches of the wall, to increase the resonance in their theatres. They were used especially in Corinth, from whence Nummius carried them to Rome. Sometimes vases of baked clay were used for cheapness. Vitruvius says that the bells were suited to certain notes of the gamut. He explains at length the way of making and placing them, as represented in the drawing by Kircher, Fig. 36. He recommends that the bells should be so constructed as to give the fourth, fifth, octave, eleventh, twelfth, and double octave, or the series of notes—

sol, doh, re, sol_2, doh_2, re_2, sol_3.

Kircher thinks this contrary to the laws of harmony, and substitutes—

sol, si, re, sol_2, si_2, re_2, sol_3,

which seems to us correct. Probably the brass bells emitted no sound, and the resonance was produced by the air contained in them and the niches of the walls. Sounding-boards in musical instruments are intended to intensify the feeble sounds emitted from the strings, whose surface is too small to put a mass of air in motion—they divide it without making it vibrate; it is necessary, therefore, to stretch them across a sounding-board, which receives the vibrations, and propagates them with more effect. A tuning-fork becomes very audible when it rests on wood. For this reason diapasons are attached to a wooden case to increase the sound; the case also causes the air in it to resound, and thus adds to the effect. It is necessary that the size of the case should be proportioned to the note to which it belongs, or the effect would not be produced.

FIG. 36.—THEATRE OF VITRUVIUS.

Elastic bodies of a certain form—sticks, chords, membranes, strings, &c.—have their own peculiar notes, which they give when struck, or which they prefer to reinforce by resonance. The volume of air contained in the case of a diapason has its own particular note, which must accord with the sound which it is capable of reinforcing. M. Helmholtz applied this principle to make an instrument for analysing sounds, called the sonorous globe (Fig. 37). It consists of a

Fig. 37.—Sonorous Globe.

hollow sphere of glass or metal, with openings at two opposite points: one of a form to receive the sound, while in the other a tube is inserted of ivory and india-rubber, to apply to one ear while the other is closed. The volume of this globe, and the size of its orifice, determine the note which it is adapted to reinforce. If that note exist in any noise, it will be heard resounding in the globe, but no other note will produce any effect. In this way, a note can be distinguished in the midst of confused sounds, which veil it entirely from the naked ear. A series of these globes, of different sizes, supplies an apparatus for analysing sounds, of much importance in acoustic researches. If two diapasons of the same note are placed even at a considerable distance, with their openings facing each other, then, if one be struck, and the sound arrested by laying the hand on it, the note will be heard from a distance carried on by the other diapason—

"Et sese lampada tradunt."

Here the vibrations are sent through the communicating column of air, the atmosphere transmitting the vibration in the air contained in one case to the other, which responds. A violin or stringed instrument will sound if the note to which it accords is given at a distance; but sounds which it does not render produce no effect on it.

Kircher mentions a large stone which vibrated to a certain organ-pipe. We have often heard of the famous pillar in a church at Rheims, which vibrates perceptibly at the sound of a bell, while all the others are immovable. Boyle asserts that he has often felt with his hand the pews in church vibrating at the sound of the organ, or at the human voice, certain notes producing more intense effects than others.

A glass may be broken by the voice. Every glass has its own note, heard when it is tapped or broken; so that if a man with a true, strong voice, pronounces the note on the edge of the glass, it will break in a few seconds. The octave of the note is said to be equally effectual. Thin convex glasses are the best for trying the experiment. The sound of a violin would answer, while the blast of a trumpet would not. A German physician saw it done in an inn by a man who made it his trade. Several glasses were ranged before him; he struck them in succession with a key to get the note, then bending down sounded the same note vigorously, and the glasses broke. There was no proof that the glasses had not been prepared: a slight scratch with a diamond would have made success more certain.

It is curious that the earliest mention of this class of facts should be in the Talmud. "It was said by Ramé, the son of Jacheskel: If a cock shall put his head into a vessel, and break it by his crowing, the owner must pay the whole price. Rabbi Joseph says, 'These are the words of the Master: If a horse by neighing, or an ass by braying, break a vessel, the owner shall pay the half of the price.'" The writers of the Talmud who invented these niceties of law must have had exuberant imaginations.

We have just seen that the phenomena of resonance are always produced by vibrations in elastic bodies. Generalising from this, we perceive that all sound results from the vibration of elastic matter, so that sound may be defined as a vibratory movement, perceptible by the ear. But before enlarging on this, we have a few words to say on the acoustics of churches, theatres, &c.—a difficult problem, which has been but little studied. How should a hall be constructed, so that the sound emanating from one point may be transmitted distinctly in all directions?

The ancients built circular amphitheatres, with the seats in raised circles, and semi-circular theatres, with the stage, not extending far back, enclosed in thick, solid walls. But the only roof was a covering to keep off the sun's rays, which, though it could not fail to reflect sound, was not taken into account by the architects. They succeeded in so disposing the seats that the actor's voice should proceed directly to all the hearers, numbering often some thousands. Even in the ruins of such theatres, we can see that this end was generally attained. Every word spoken in the arena can be heard at the farthest seats. The theatre of Hadrian's villa at Tivoli, the circus of Murviedro, and the amphitheatre at Nismes (Fig. 38) are remarkable in this respect.

The only means employed by ancient architects to augment sound were the vases or bells already mentioned.

Fig. 38.—Amphitheatre of Nismes.

Public affairs were transacted in an open building called a forum. Under the blue heavens they enjoyed their amusements, took counsel, and listened to harangues. But now that civilisation has left its cradle to find a home under ruder skies, for this simple architecture various kinds of halls, circuses, concert-rooms, theatres, houses of legislation, &c., are substituted. Platforms, pillars, stalls, boxes, pews, introduce great difficulties into the propagation of sound, by their powers of resonance and reflection. We must proceed on a new plan to discover the method of applying the science of acoustics to modern buildings. Domes are generally unfortunate in their effect; they produce a too powerful and too prolonged resonance. Under the dome of St. Paul's the sound seems to run along the walls. In the Rotunda at Rome this resonance produces such singular effects, that it is said many people go to church for the sake of hearing them. In the circular concert-room of the Fine Arts Society in Berlin, where the walls are broken by a large number of deep embrasures,

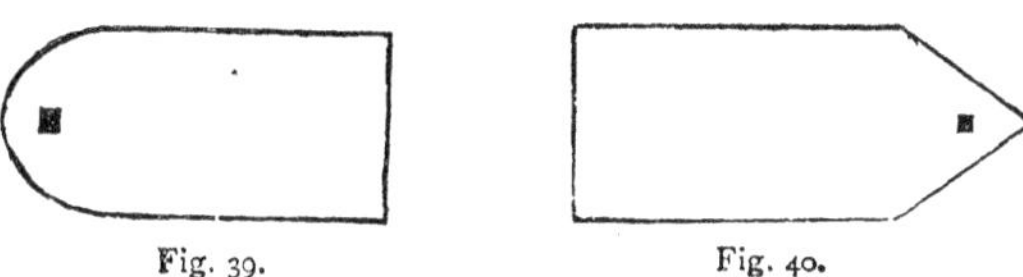

Fig. 39. Fig. 40.

this inconvenience is not met with. The dome of St. Mary's at Dresden is remarkable for the same absence of resonance.

There is no advantage in elliptic arches or halls, the ellipse only serving to concentrate sound at a particular point. The parabola, which makes diverging rays parallel, has some recommendation. The speaker's desk should be at the focus of the curve. Chladni proposes to terminate a rectangular hall with a parabola. This arrangement (Fig. 39) is found in some ancient basilicas. It might be completed by giving a parabolic form to the roof over the platform. A sounding-board of this nature is sometimes placed over the pulpit; its mode of action is the same as that of the apparatus for reflecting in lighthouses. In a concert-room or hall it might be an advantage to construct over the platform a spherical dome, with its axis directed to the centre of the hall. Another idea of Chladni's is to place the speaker's platform in a semi-conical space at the extremity of the hall (Fig. 40), but he admits that this arrangement would be unsightly and difficult of construction. In theatres, of course, no reflector could be placed behind the actors. The only suggestion deserving attention is to employ, like the ancients, triangular columns turning on their axes, instead of the folding screens through which so much sound is lost (Fig. 41). The arrangement of the seats in a semi-circular form would not suit the exigencies of the

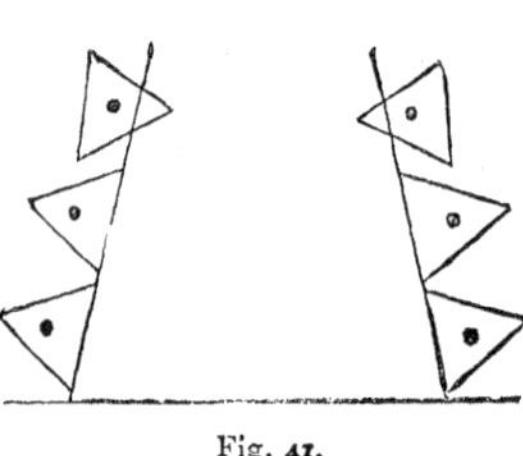
Fig. 41.

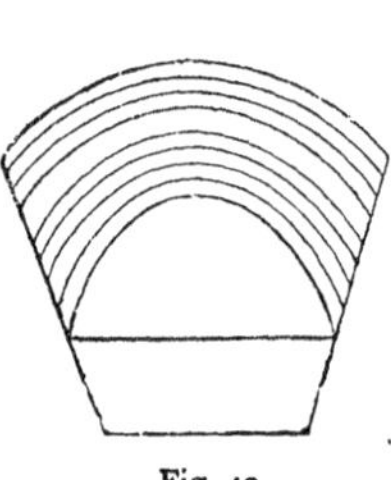

Fig. 42. Fig. 43.

modern drama. An advantageous form is given in Fig. 42. The theatre of Parma, which is celebrated for its acoustic properties, is given in Fig. 43. The boxes in front of the stage are the great defect in modern theatres. Zamminer compares them to monster traps for strangling sound

Unfortunately the architect is compelled to consult the wishes of those who come not to hear, but to see.

In the construction of our churches and amphitheatres, the simplest laws of acoustics are neglected, and consequently very imperfect effects obtained. The commonest defect is an excessive sonorousness, which prevents words from being distinctly perceived. The semi-circular room of the Fine Arts School in Paris, though beautifully decorated, is miserable in this respect. The great Amphitheatre of Physics and Chemistry in the Jardin des Plantes, and the Amphitheatre of Physics in the College of France, are inconveniently sonorous. They have tried to remedy it by using drapery to deaden the walls, and pieces of wood to impede the vibrations of the raised seats; but this modification is of little use. In the church of St. Paul, Boston, which has the same defect, the preacher's voice can only be heard once a year, on Christmas Day, when it is decorated in such a way that the arches are less sonorous.

The semi-circular form, so often given to amphitheatres, produces great inequality between the seats at the centre and those at the extremities, as in the Amphitheatre of Physics of the Sorbonne, and that of the Conservatory of Arts and Trades, but there the inconvenience is lessened by the chair being differently placed. The most advantageous form is that approaching the quarter of a circle, because the walls direct the sound to the hearers.

For placing the raised seats, the general rule is to follow a line direct from the platform to the beginning of the roof. A concave curve would be more advantageous, as it would obviously allow of the back rows hearing better. Mr. Scott Russell, M. Lacheze, and others have proposed several curves for this purpose.

The most original project for improving the acoustics of theatres is that suggested to Chladni by Langhaus, of Berlin. He would direct from the stage to the spectators a slight current of air, which should carry the words of the actors. It would be produced by skilful ventilation.

CHAPTER VIII.

SOUND IS A VIBRATION.

Trevelyan's Instrument—Singing Flames—Pendulum—Undulations of Water—Progressive and Stationary Waves—Vibration of Rods, Strings, Boards, and Tubes—Graphic Method.

UP to the present we have only considered the phenomena of sound as affecting the senses. It is now time to consider what produces them. The phenomena of resonance point to the conclusion that sound can only originate in the vibrations of a ponderable body.

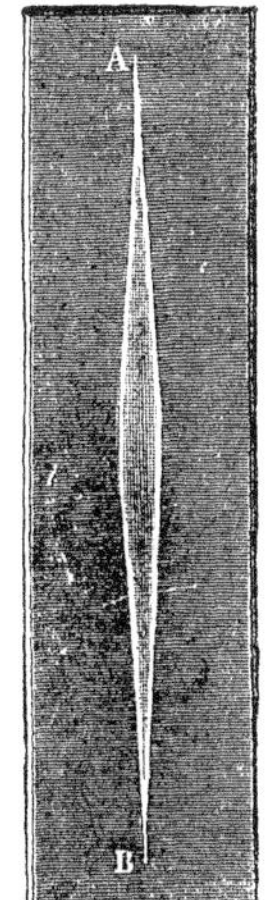

Fig. 44.

Common experience shows us that a sound of any force is always accompanied by vibrations perceptible to the touch. Drums beaten in the streets shake the window-panes. The report of a cannon makes the earth tremble; those near enough feel a shock in the chest. In a concert-room, turning the opening of a hat toward the orchestra, you may feel the trembling of the air by placing your finger-ends on the crown. In many cases it is easily proved that sound cannot be produced without a concurrent vibratory movement. A stretched chord when struck makes oscillations which are visible, and, owing to the persistence of luminous impressions, it takes the form of Fig. 44.

The outline of a diapason becomes indistinct while it sounds, because of the rapid motion. A glass bell, rung by means of a fiddle-stick or wooden hammer, will communicate violent shocks to a little ivory ball hung beside it. Each time that it touches the bell, the ball is thrown away, returning again and again as by an irresistible impulse, only to rebound once more. If the edge of the bell be touched with a pencil, it grates against the vibrating glass; or if a horizontal bar of steel be rubbed between the thumb and fore-finger with a little colophony, it gives a sharp sound; and if the ivory pendulum touch either extremity, it rebounds with great force.

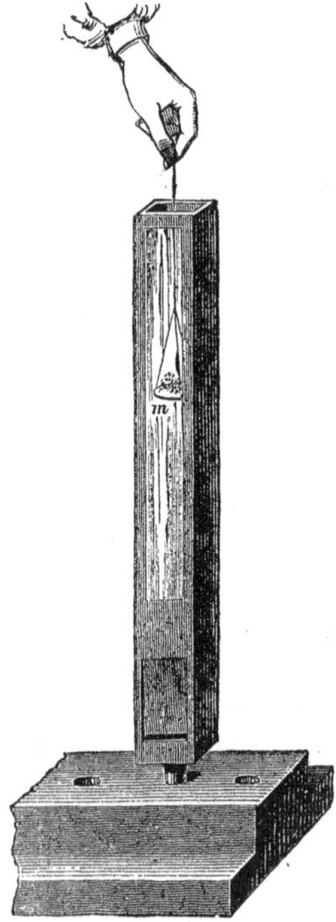

Fig. 45.

Plates of brass, wood, or glass give different sounds, according to the manner of striking them. Sand sprinkled on the surface assumes regular curves, marking the lines of repose. A membrane stretched upon a cardboard frame, and hung by three threads in the pipe of an organ, will throw to a distance the powder scattered upon it. The better to show this, a glass pipe can be used for the organ (Fig. 45).

It is always easy to produce sounds by mechanical action repeated at short intervals. The buzzing of a fly's wings, the chirp of a cicada or grasshopper, are examples of this kind of sound. A flexible card pressed against the edge of a cog-wheel makes a sound, which grows sharper as the rotation becomes more rapid; this is the principle of the rattle. In the siren (an apparatus that will be described further on) a stream of air or liquid is directed against a perforated revolving disc; this stream either passes or is arrested alternately, thus giving birth to a remarkable sound. In the reed stop of an organ the sound is produced by the vibrations of an elastic tongue. The lips tremble while playing on the oboe or clarinet.

It sometimes seems as though sound might be produced by a continuous movement; the flute and the common whistle seem influenced by the action of an uninterrupted current of air. But in these cases the current is broken and divided into two streams by the orifice, one part entering the mouth-piece, the other escaping into the outer air. The current first compresses that portion of the air next the orifice; this latter, reacting by its elasticity, resists the current, then gives place again, repeatedly; so there is, in reality, a continued series of vibrations. Wertheim succeeded in playing upon pipes submerged in liquid, by the injection of a stream of the same liquid. The sounds he obtained had the same musical character as when the pipes were played upon by air. Cagniard de Latour had previously to this made glass tubes vibrate in water by means of friction, so that the water became sonorous.

We must here mention the "rocker" of Trevelyan, in which the sound results from the contact of two metals unequally heated.

In 1805 M. Schwartz, inspector of one of the foundries of Saxony, having placed a silver cup, still hot, upon a cold anvil, heard, to his great astonishment, musical sounds coming from the metal. Professor Gilbert, of Berlin, repeated this experiment, and described how the cup vibrated so long as the sound was heard, but grew quiet as it cooled and the sound ceased; he did not attempt to explain the phenomenon.

About 1829 Mr. Arthur Trevelyan, wishing to melt resin with an iron, found the iron was too hot, and laid it against a block of lead to cool. Scarcely had the iron touched the lead, when a sharp note was heard coming from it, something like a Northumberland flute; at the same time he saw the iron moving in rapid vibration. Mr. Trevelyan set himself then to study these facts, and he gave an explanation of them which seems to be the true one. The vibrations he supposes to be caused by the sudden expansion of a cold body when brought into contact with a warmer. At the moment when the hot iron touches the lead at a given point, the lead expands and repulses the iron; the iron then touches at some other point, where the same thing occurs, whilst the point first touched cools and contracts. By this play of alternate expansion and contraction the "rocker" is able to produce music. It is usually made in brass, of a prismatic bar, the lower angle having a hollow groove. This is fixed on a round handle. When heated to about the temperature of boiling water, or a little more, it is placed on a piece of lead. Mr. Tyndall made the same experiment with a heated shovel, which he balanced on two pieces of sheet lead fixed in a vice. It immediately took a see-saw motion, and gave out a musical sound, which could be modified by lightly touching the handle.

Sometimes a musical vibration may be obtained by a simple coin or ring laid upon a piece of lead, after having been sufficiently heated.

When a current of air is heated and cooled periodically at a certain point, there results a succession of alternate dilatations and contractions, which may prove a source of sonorous vibrations. This is illustrated by the apparatus of Fig. 46. It is composed of a glass tube, in which is fixed a small metallic web. This is heated red-hot by a spirit-lamp. After a few moments a plaintive sound—a sort of low moaning—seems to float around the tube; gradually it swells, increases, becomes very loud; then, as the web cools, the sound dies away, and the tube becomes silent again. The sound is caused by the ascending current of air becoming heated as it passes through the web, and cooling as it leaves it. Indeed, by lowering the tube towards a horizontal position it may be stopped momentarily, because of the interruption of the current of air. The mysterious sounds which were heard to proceed from the statue of Memnon at sunrise were, very probably, caused by the currents of air in the hollows of the stone being heated by the sun's rays (Fig. 47).

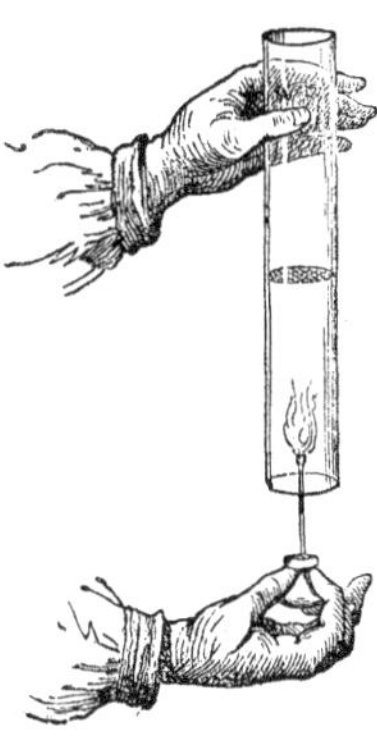
Fig. 46.

We often hear the gas sing when the jet is stopped by an obstacle which prevents the free passage of the current. The jet, instead of being continuous, is intermittent, and the gas escapes by pulsations. A current of hydrogen in a glass tube would produce the same effect. This little circumstance has given rise to a number of beautiful experiments by Count Schaffgotsch and others. Introduced into a glass tube is a small brass burner, with a gas flame (Fig. 48).

If then a note be sounded at a distance, in harmony with the glass tube, the air within begins to vibrate, and communicates its pulsation to the flame, which grows tall, and trembles, and begins to sing in its turn. It may be silenced by

Fig. 47.—Statue of Memnon.

pressing a finger on the opening of the tube, but will sound again for another call of the voice; only the true note must be produced, or the flame will not respond. With four flames and four tubes, a little organ may be made to give the chord doh, mi, sol, doh, in perfect harmony, whose music is sustained as long as the flame continues to burn. Sometimes, too, it will happen that the flame will begin to sing spontaneously, if its point be placed at a certain part of the tube.

It is easily proved that the sound of singing flames is produced by a pulsation of gas burning in the tube. The flame changes alternately from yellow to blue, according to the quantity of gas which comes to feed it. If the head be moved quickly from right to left, the flame will seem to separate into a number of blue and white images, which being received on different points of the retina, are not confused in the eye. The result may be better obtained by using an opera glass during the experiment. The best means of separating the successive appearances of the flame is, however, furnished by the revolving mirror. This is a mirror with two, three, or four faces, rotating round a vertical axle. It causes the flame to appear every moment in a new direction, the result of which is a kind of luminous ribbon, *continuous* so long as the flame remains still, but breaking into a chaplet of brilliant pearls when it begins to vibrate. There is a succession of little stars, followed by luminous trails of a rich blue, such as we see in jets of gas when the wind blows on them. These trails terminate in spaces of complete darkness, which seem to indicate that the flame is momentarily extinguished, though immediately rekindled.

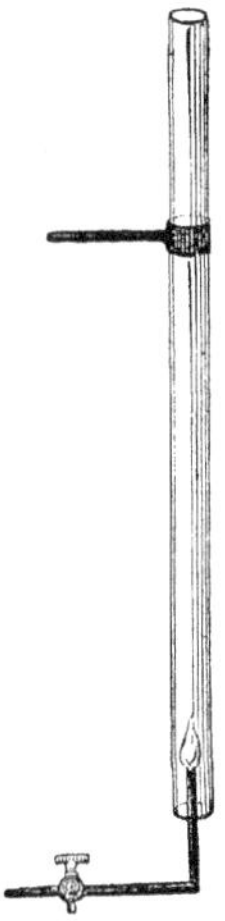

Fig. 48.

Sonorous flames may also be studied by means of a revolving disc, perforated with a circular row of holes. A vibrating body, looked at through such an apparatus (called a stroboscope), appears to move with diminished velocity. It is as if we had a microscope to magnify time. The vibratory movement is a motion of going and coming, reproduced at equal intervals, in a uniform rhythm. The oscillations of the pendulum give a curious example of this Moved from its position of repose, the pendulum immediately returns because of its weight. It falls, but in falling it acquires an increase of velocity, and passes the starting-point. It mounts to an equal height on the opposite side. It cannot mount higher, for the weight draws it back while it swings, thus gradually destroying its velocity, which becomes nothing as at the moment of first setting it off. Then the pendulum is found exactly in the same condition as at first; and the action recommences in an opposite manner: it descends, passing the point of equilibrium at its maximum speed, and returning to its starting-point with no velocity. Thus it has accomplished a complete oscillation, going and returning, or two simple oscillations in a contrary direction. Should nothing stop it, it will continue indefinitely to move thus from side to side of its vertical; but the resistance of the air, and the friction of the thread at the point of suspension, together with other causes, diminish by degrees the scope of the oscillations, and so bring the pendulum at last to rest. It is ascertained that all oscillations are accomplished in a definite time. A pendulum a little over a yard long performs one oscillation in a second.

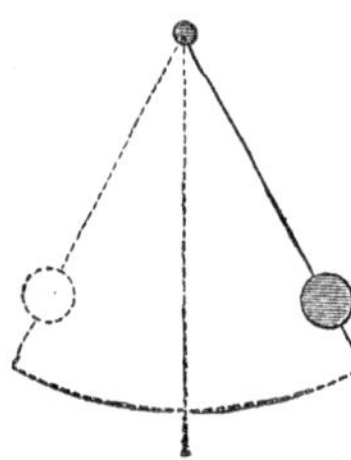
Fig. 49.—The Pendulum.

The motion of the pendulum is kept up by the force of gravitation. The vibrations of a sonorous body are usually sustained by the force of elasticity. Like the vibrations of a pendulum, they are finally extinguished by the action of different resisting forces which are constantly tending to destroy them. The duration of the vibration of perceptible sounds varies from the tenth to the twenty-thousandth part of a second.

As to the particular nature of these vibratory movements, they may be of different kinds. In the air they form alternate condensations and dilatations. A prismatic body may contract and dilate lengthwise, or bend transversely, or even perform rotatory vibrations.

When sound is propagated the vibrating air-particles do not sensibly change their place, but only move near their positions of rest for a short space, and the motion or pulse only is transmitted to a distance. Therefore it is that water scarcely seems to be displaced when traversed by an ordinary wave. To prove this, throw a stone into a piece of quiet water. Around the point of commotion we see concentric rings, which are propagated to the shore, describing larger and larger circles. On their way they meet with many floating bodies—pieces of wood, withered leaves, and straws. Light as they are, these are not carried away. We see them rise at the approach of the wave, and sink as it passes, but they do not perceptibly change their place. It is not then a material wave which is carried on the surface of the water; that which appears to be carried is merely the shock or impulse, and the deformation that results from it. The rings dissolve each moment, and each moment are formed anew, with fresh particles, which in their turn quickly come to repose. Let us now imagine, instead of a single stone, a number thrown in one after another, at regular intervals, to the same place; the waves that they excite will also break upon the shore at regular intervals, but they will not carry the particles of water very far; they mount and fall continually, and pass on the impulse they have themselves received.

The interesting experiments of Ernest Henry and William Weber showed that liquid particles generally move in

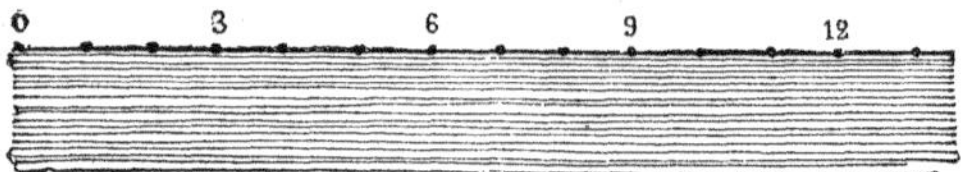

Fig. 50.—Undulations of Water.

circles, while the wave travels onwards. To make this plain, let us suppose that each particle makes a complete circle in the time that the wave takes to go from the point o to the point 12 in Fig. 50: it will make the twelfth part of a circle as the wave clears each of the twelve spaces

Fig. 51.—The Quarter of an Undulation.

between the points o and 12. At the moment the wave touches point 3 (Fig. 51) the particle o will already have had time to accomplish three-twelfths or one-quarter of its circle; the particle 1 two-twelfths or one-sixth, and the particle 2 one-twelfth of the circle, while particle 3 will scarcely have begun its movement. At this moment the particle o will have reached the lowest point of its course, and then will begin to mount the opposite side.

The next figure (Fig. 52) represents the situation of the particles by the time the wave has reached point 6. The particle o has finished the half-circle, particle 3 a quarter, and so on. It is now 3 which is at the base of its path,

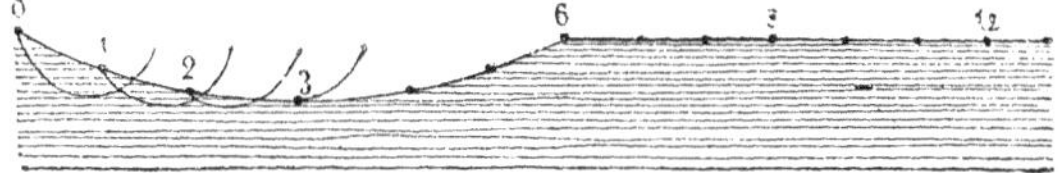

Fig. 52.—Half of an Undulation.

whilst o is again on the general level. Between o and 6 is a hollow.

In Fig. 53 the first particle has described three-quarters of a circle, and is seen on the culminating point of its course; the particle 3 has made half its journey, and re-gained its first level; the whole set from 3 to 9 form a

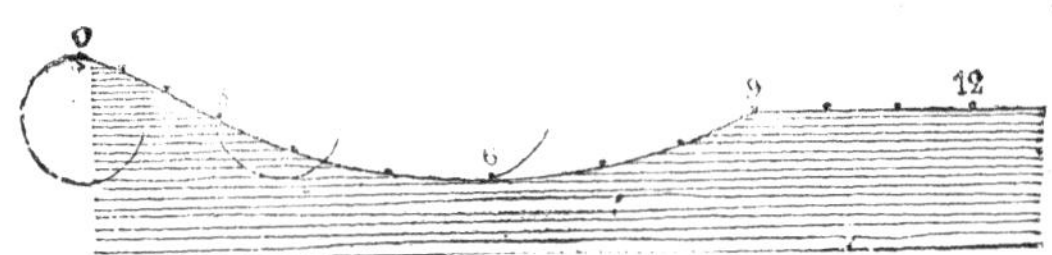

Fig. 53.—Three-quarters of an Undulation.

hollow undulation, just as the set between o and 6 did formerly.

Finally, in Fig. 54 the hollow is displaced for three points—that is, from 6 to 12. The point 3 is now at the summit of its course; while point o, having described an entire circle, has returned to its primary position. Between o and 6 there is a crest. This elevation, and the depression which extends from 6 to 12, taken together form an entire wave, and the interval it fills is called the wave-length. It will be noticed that at the depth of the hollow the particles are at a distance from one another, while towards the crest of the wave they are close together. The same thing is

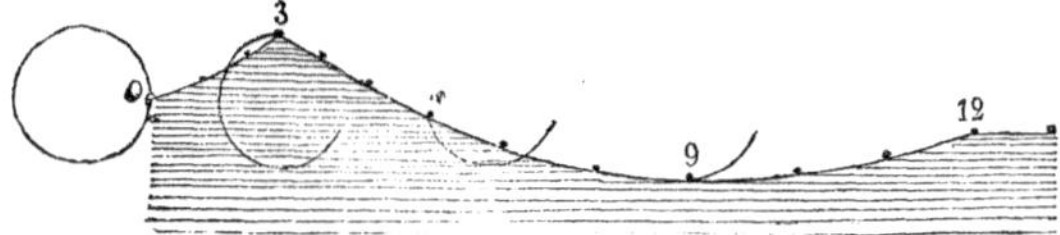

Fig. 54.—Complete Undulation.

repeated at regular intervals afterwards. When the particle o has finished its second revolution, the particle 12 has only accomplished its first; there is one complete wave between o and 12, and another between 12 and 24 (Fig. 55). When the particle o has made three turns the waves are propagated up to the point 36; when it has made four turns the

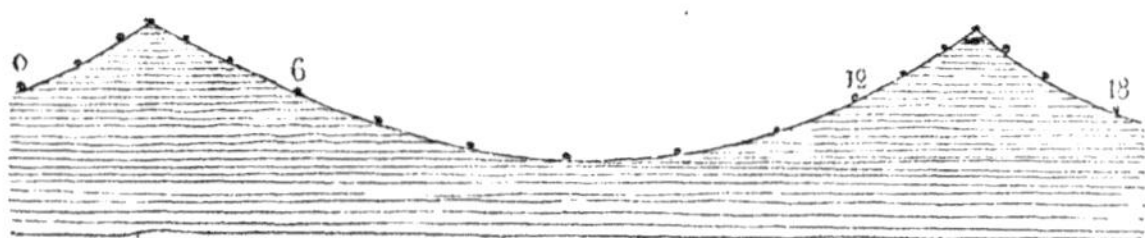

Fig. 55.

waves have reached point 48, and so on, advancing a wave-length at each oscillation.

Particles may travel in ellipses instead of circles, and these ellipses may become so elongated as to be transformed into straight lines. Then the liquid particles only rise and fall vertically; they simply make transverse vibrations, as we may see them do in chords, metal plates, and membranes.

The general form of the wave remains the same, but the trough and the crest become symmetrical, the one being always the reverse of the other, as is shown in the following curves (Fig. 56), which represent the progress of a transverse vibration. Such are the undulations of the ether which produce light.

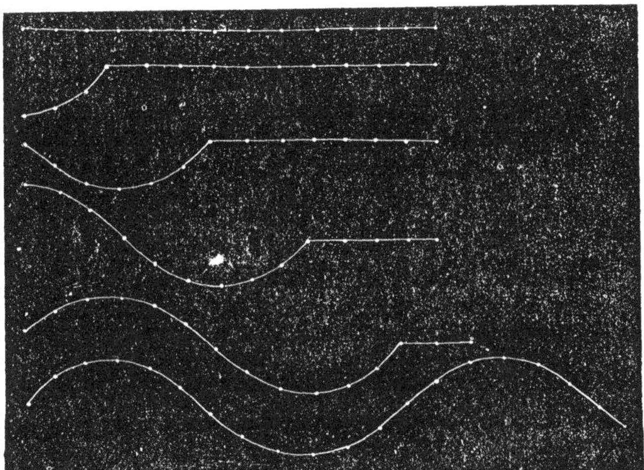

Fig. 56.—Progression of a Transverse Vibration.

If the orbits of the particles, instead of becoming vertical lines, changed into horizontal lines (the propagation of the wave being always supposed horizontal), we should have longitudinal vibrations, analogous to those of gaseous bodies. The particles then can only separate and approach by turns, whence result alternate dilatations and compressions, as may be seen in the curves in Fig. 57, which represent the progression of a longitudinal wave.

In a body of cylindrical form another class of vibrations may be seen—tortuous or revolving vibrations. The particles circulate round the axis of the cylinder, and the motion is propagated in the same manner as in other cases. Each particle begins its excursion a little after the preceding one, and therefore remains a little behind it, in all the phases of the oscillations, which they pass through together.

Fig. 57.—Progression of a Longitudinal Vibration.

In this way the progressive waves are propagated in an unlimited medium. Thus sound is transmitted in the open air, light through the ether, and undulations in an unbounded sheet of water. We observe these waves to move along as if each phase of the movement of the first particle were transmitted successively to all the file. In transverse vibrations we see the summit of the wave displaced, and travelling along the chord. In longitudinal vibrations it is the compressions and dilatations which are transmitted (Fig. 57). An india-rubber tube, fixed at one end and held by the hand at the other, is shown in Fig. 56. A slight stroke at one end will send a transverse wave undulating along the tube, thus forming the curve. It may be followed with another wave, by striking again on the end of the tube the moment it becomes still; then with a third, and a fourth, and so on, till the first has reached the wall against which the tube is fixed. From this instant the phenomenon changes its aspect; the waves being unable to advance are obliged to return,

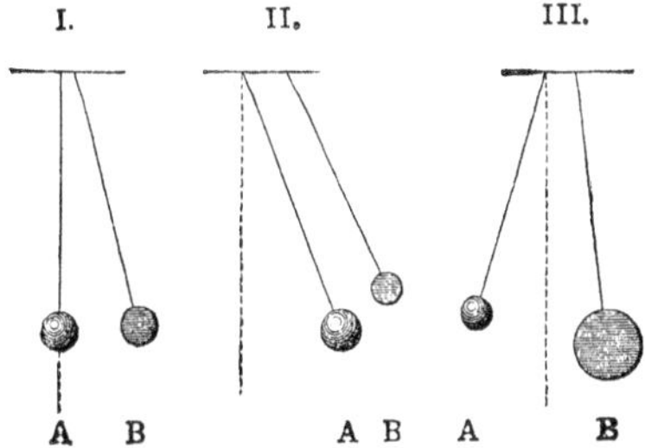

Fig. 58.—Shock of Elastic Balls.

and the returning waves meet the later, which are still advancing; hence the result known as "fixed waves.'

The fixed waves characterise the sonorous vibrations of elastic bodies, whether they give out their own sounds, or only resound under the influence of repeated shocks. They can be easily distinguished from progressive waves. In the one the particles vibrate one after another, whilst in the fixed waves they vibrate altogether. These waves do not travel; they are born, live, die, and rise again always in the same place.

This change is owing to the intervention of reflected waves. The laws which govern these phenomena are complicated enough. To give an idea of them, let us consider what happens at the meeting of two elastic bodies. Suppose A and B (Fig. 58) to be two billiard-balls hung by two parallel threads. Raise the ball A, and let it fall against ball B; if their size be equal (I.), A will remain in repose after the shock, giving up all its velocity to B, and B will be thrown forwards. If the ball A be larger than B (II.), it will pass the vertical line with a velocity scarcely diminished, chasing the smaller ball before it. Finally, if A be smaller than B (III.), it will be thrown back with more or less force; the greater the resistance opposed by the mass B, the stronger will be the rebound.

The same thing takes place when a vibration is propagated in an elastic medium. The balls A, B, in Fig. 58 (I.), represent two neighbouring particles which transmit a progressive wave. B receives all the velocity from A, and A remains in repose till another impulse comes to disturb it. But if A and B are, so to say, the bordering columns of two media of differing density, we fall into one of the two cases represented by II. and III. If, for example, the medium B be less resistant than the medium A, the particle A will pass forward while communicating its velocity to particle B (II.). If, on the contrary, the second medium be more resistant than the first—if, for example, B represent a fixed obstacle—the particle A will be thrown back, and B will be scarcely stirred.

Now, in these cases what must follow? The particle A not being at rest will become a source of movement for all the particles behind. The result will be a reflected wave, which will carry back the movement given by A, either in the direction that A was pursuing before the shock (II.), or in the contrary direction (III.).

These comparisons will serve to give an approximate idea of the phenomena accompanying the reflection of a sonorous wave. The first case (II.) represents the reflection of a sound in the interior of a solid body which vibrates in the air, A being a point of the surface, and B a particle of air.

A reflection of the same nature takes place at the extremity of a tube filled with air, opening into the atmosphere; for the surrounding air, because it moves more freely, has less resistance than the air inside. Therefore the sound which comes from an open tube is partially reflected by the surrounding air, and returns to the tube. This result, indicated by theory, may be verified by experiment: at the end of a very long open tube a faint echo is formed. Biot observed that when he spoke at one end of the water-pipes of the aqueduct at Arcueil, the sound was echoed back to him six times.

The case shown in Fig. 58 (III.) is the same as we have in fixed obstacles. Sound is reflected in this manner, in the interior of a closed tube, from one end to the other. A simple apparatus, which we have not time to notice now, would show how, in either case, the direct waves and the reflected combine so as to produce fixed waves, separated by points of repose, called nodes.

The particles comprised between two consecutive nodes form what is called a simple wave.* Agitated by a common motion, they all rush forward in the same direction, and return in a contrary one. The centre of each wave is also a *centre of vibration*. There the commotion is at its

* The simple wave is equivalent to the half of a complete or double wave, just as a simple vibration is the half of a complete or double vibration.

maximum ; from the centre to the nodes it diminishes. the extent of the excursions decreases, and at the nodes all movement has ceased.

The particles of two consecutive waves always vibrate in opposite directions. If they rise in one, they sink in the other, and *vice versâ* (Fig. 59) ; if on the one side they

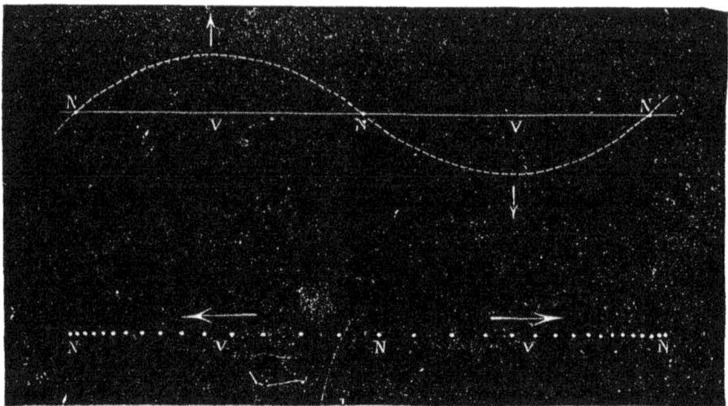

Fig. 59.—Nodes and Centres.

approach or depart from the node that separates the two waves, they approach or depart equally on the other side.

The interval between two nodes or two centres is a simple wave-length, which is half an entire wave-length. The length of a fixed wave is equal to that of a progressive wave ; it is the measure of the advance made during the time a single vibration lasts ; in other words, it is the space traversed by sound during one vibration.* Thus when a

* A simple wave-length corresponds to a simple vibration, as a double or entire wave-length corresponds to a double or complete vibration. Sometimes one and sometimes another of these quantities is employed, therefore it is necessary not to confuse them.

vibration lasts the millionth part of a second, the corresponding wave-length is thirteen inches if the sound be propagated in the air, and fifty-six in water, &c., since these numbers represent the spaces traversed in the different media during the millionth of a second.

In the reflection from a fixed obstacle, a node is formed close against it, since the direct and the reflected shock, being in contrary directions, neutralise one another. Nodes are therefore found at the points of suspension of a vibrating body—at the ends of a string, for instance, or the points where a metal plate is held in a vice. The position of the other nodes depends on the shape of the sonorous body, and the sound given out by it.

Any elastic body will generally return all the sounds which meet it, but the resonance varies greatly in intensity. It is strong only when the nodes of the fixed waves, resulting from the interior reflection of sound, follow certain regular directions, and in this case they continue after the producing cause has ceased to act. The sounds which develop such a peculiar resonance in a body are such as are produced by a mechanical shock—in other words, they are the sounds properly belonging to a body. Any other sound finds but a feeble echo.

Let us now consider the fixed vibrations of some sonorous bodies, and find out the arrangement of the nodes which characterise their specific sounds. Take first a string, fastened at both ends. In this case there is a node at either end, since the extremities are motionless; there may be, besides, any number of nodes at intervals from one end of the string to the other. If it vibrates transversely in all directions, all its points will simultaneously describe the same kind of orbits, but of different dimensions, that of the centre of the chord being the widest. This orbit may be a right line, vertical or horizontal, an ellipse, a circle, or any other curve, according to the mode employed to produce the vibrations. If it be a right line, the string will vibrate in a plane; if it be a circle, it will form a spindle (Fig. 60). To make it vibrate with three nodes, we have only to touch the middle of the string lightly with the finger while striking one of the two halves with the bow; the string then divides into two conical spindles or segments, separated at C by a node, and vibrating in contrary directions (Fig. 61). Touching it in the same way at different points, we may obtain three, four, or even five segments of the string, in each case giving a different sound according to the manner in which it is divided. The immobility of the starting-points may be demonstrated by placing slips of paper upon them, which will remain perfectly quiet while on the nodes, but at any other point will be thrown off immediately (Fig. 62).

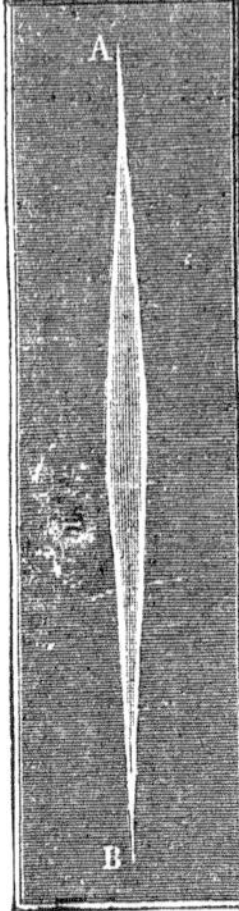

Fig. 60.

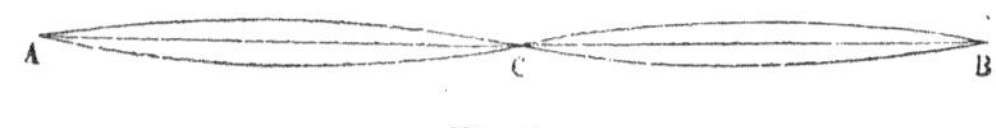

Fig. 61.

By rubbing the chord lengthways with a little resin Fig. 63), the longitudinal vibrations are shown, consisting of alternate dilatations and contractions. When there are only two nodes at the extremities A, B, the section A, C, dilates, while B, C, contracts, and *vice versâ*; the middle

C becomes a centre of vibrations where the movement of translation is a maximum, but where the density remains the same. In the nodes A, B, the density, on the contrary, changes most, and there is no translation. It could not possibly be otherwise, for since those particles at C move

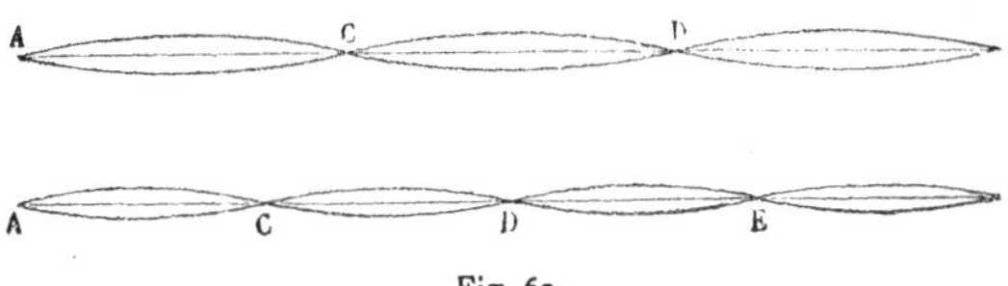

Fig. 62.

more than the others they will trench upon those in front, forcing them to a compression; at the same time distancing those behind, which, consequently, must separate more and more.

Now the chord may be again subdivided into portions of an equal length, separated by nodes which will become the

Fig. 63.

centres of successive compression and dilatation. From the two sides of each node the particles move in contrary directions; compression takes place when the node becomes the meeting-place of two files, and dilatation when it is the starting-point of two files moving away again (Fig. 64).

It often happens that a string is stirred at the same time by longitudinal and transverse vibrations, more or less complicated, to which may be also added rotating vibrations.* Each particle then describes an orbit in the form of a spiral slightly distorted. If you picture a poor fiddle-string tortured by the bow of the fiddler, who strokes it and strikes it, pinches and stretches it by turns, you will

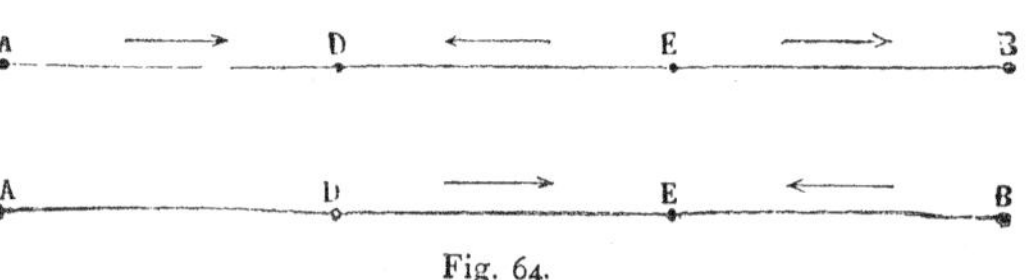

Fig. 64.

not marvel to see it execute curves such as no geometer has ever dreamed of.

To get transverse vibrations from a prismatic metal plate, it may be either fixed by one end or laid upon two triangular wedges (Fig. 65). A series of centres and nodes

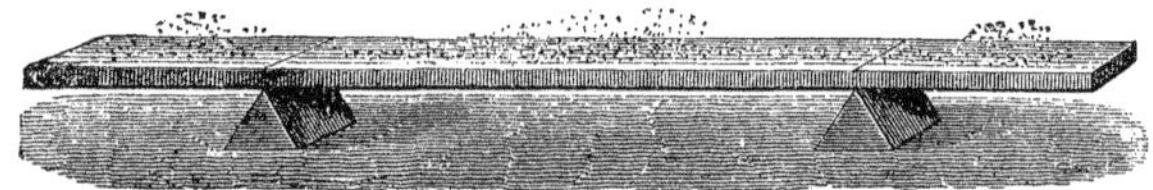

Fig. 65.

will then be seen, whose distribution depends on the manner in which the rod is supported. The general rule is that there are always centres at the free extremities, and nodes at the fixed points. The nodes are shown under the form of straight lines, which cross the plate, and which

* A chord cannot vibrate transversely without lengthening slightly, and this occasions longitudinal vibrations. This longitudinal sound is sometimes recognisable in the *la* of the violoncello.

may be rendered visible by throwing sand upon the plate while it vibrates. The grains of sand unable to remain at the centres, where the tumult is at its height, take refuge at the nodes, which afford them a quiet asylum, and group themselves in fine right lines, called the lines of repose, or nodal lines.

Tuning-forks belong to the same category as the prismatic metal plates; they vibrate so that there are two centres at the extremities of the branches, which alternately approach and separate, two nodes close to the base, and a third centre in the midd'e of the fork. This lower centre makes the stem rise and fall, so that when placed upon a wooden table it causes it to resound by the incessant vibration (Fig. 66)

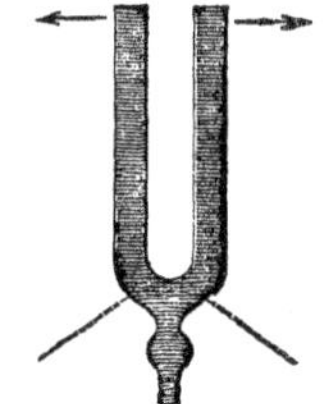

Fig. 66.

The longitudinal vibrations of prismatic or cylindrical bars develop a wonderful force. Savart, having secured a steel rod (Fig. 67), placed a spherometer opposite the free end, not touching it when at rest, but near enough to be struck at each vibration. The shocks were heard when the spherometer was at a distance of $\frac{24}{1000}$ of an inch; the total variation in the length of the rod (dilatation and contraction) being then at least double, or equal to $\frac{1}{20}$ of an inch. It would have needed a weight of 3,740 pounds, hung to the end of the rod, to lengthen it to this extent. This proves that during its longitudinal vibrations, a steel wire is subject to a traction which might become strong enough to break it. Thus, when a weight is not sufficient to break a metal

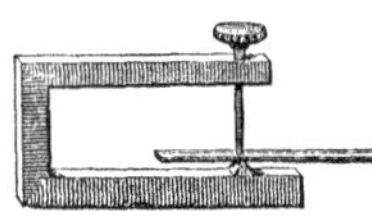

Fig. 67.

wire, or even to get a permanent elongation, either of these results may be obtained by making the wire vibrate

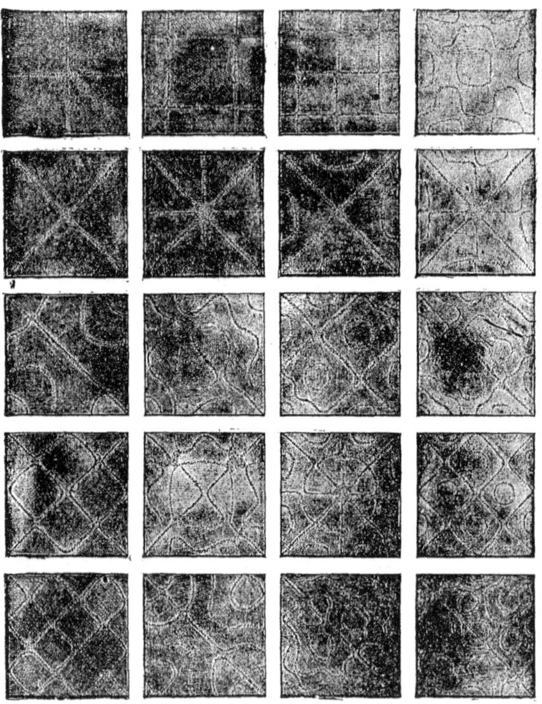

Fig. 68.

throughout its length while the weight hangs from it. For this reason it is always necessary to avoid a regular oscillation of the chains of a suspension bridge. In America, and other countries where there are great suspension bridges, they forbid regiments of soldiers walking in time, or even herds of cattle, to pass, fearing the effect of the vibration on the chains.

To make a thin plate of metal, wood, or glass vibrate transversely, the edge should be struck with a bow. The simplest means of holding it during this operation is to take

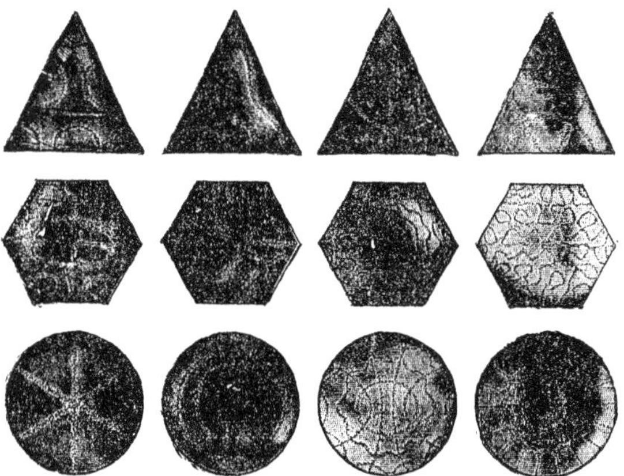

Fig. 69.—Chladni's Figures.

it between the thumb and the fore-finger, if it be small enough, or to let it rest on three fingers. The best way, however, is to fix it with four screws covered with cork, at four points, through which the nodes will pass. The bow is then drawn vertically across the edge of the plate.

If the plate be previously, or during the vibrations, sprinkled with fine dry sand, the grains of sand will be seen first to dance tumultuously, and at last to range themselves in regular and symmetrical figures. The nodal lines on the plate mark the places where there is no vibration. Each line separates two vibrating segments, where the vibrations are opposite, the surface falling in one while it rises in the other. Figs. 68 and 69 represent some of the nodal lines which may be seen on plates of different forms—square, triangular, circular, &c.

Fig. 70.—Chladni.

These beautiful phenomena were discovered and published by Ernest Florens Frederic Chladni, Doctor of Philosophy, in 1767. He passed the greater part of his life in illustrating acoustics in the different towns of Germany, France, and Italy, wherever his erratic humour led him. To him we are also indebted for the first catalogue of aëroites, and the earliest affirmation of their ex-terrestrial formation. Chladni's figures long puzzled the philosophers, who looked upon them as an unanswerable enigma. Savart endeavoured to explain them, but as usual he only involved the matter in deeper obscurity. The only useful discovery which he contributed was one made by his assistant, of using a powder of heliotrope in place of sand, and laying a sheet of damp paper over the figures, by which means they may be printed and kept for reference.

Bells, discs, and glasses vibrate with nodal lines which divide the surface like seams. If the bell or glass be turned mouth upwards and filled with water, these vibrations will express themselves in beautiful ripples upon the surface. On pouring the water in, it will be thrown away from the vibrating segments, and remain motionless in contact with the nodes. The nodes may also be discovered by suspending a little ball by a string, and letting it gently touch the vibrating surface; when the ball remains still we may know that it is on a nodal line.

The same experiments may be shown on a drum, or a sheet of paper or collodion stretched upon a frame. Owing to its flexibility, a thin membrane will easily resound under the impression of any sound whatever. The tympanum of the ear affords a striking instance of this. Therefore we may ascertain the position of the nodes and vibrating segments in a vibrating column of air by the ear, or by a little drum covered with sand.

We have already said that the vibrations of the air are longitudinal. In the vibrating segments there is agitation; n the nodes, complete repose; with alternate compression nd dilatation. The motion of the air in the segments may be communicated to a membrane, if it be struck perpendicularly; the compressions and dilatations that take place in the nodes will cause it to vibrate, if they act on one side only. The ear is especially sensitive to the changes o density in the nodes.

The flames of Kœnig (noticed more fully hereafter) allow us to make use of this property belonging to membranes, to exhibit the changes in the density of the air. These flames are supplied by a stream of gas, vibrating under the pressure of a membrane inserted in the pipe. Observed in a revolving mirror they have the appearance of a rov

Fig. 71.

of tongues, separated by black spaces (Fig. 71), whic depend upon the nature of the sonorous vibrations. A admirable means of studying the vibrations of sonoro bodies is afforded by the "phonography" first conceived b William Weber. Imagine a pendulum ending in a poir swinging exactly over a sheet of paper blackened wit smoke. Evidently, the point will clear a white line f itself through the black powder, in which it will pass fro right to left, and from left to right. But if the paper b drawn slowly back, it will touch a different point ea moment, and, instead of a straight line, there will app an undulating curve.

The same result may be obtained by using a vibrati

rod in place of the pendulum, which shall mark its way upon a piece of smoked glass. If the tube have a fine and flexible point it will trace every vibration by a zigzag on the glass. It is still better to use a rotating cylinder for this purpose, with a sheet of blackened paper fastened to it. When the tracing is finished, the paper is taken off and steeped in alcohol, which fixes the pattern.

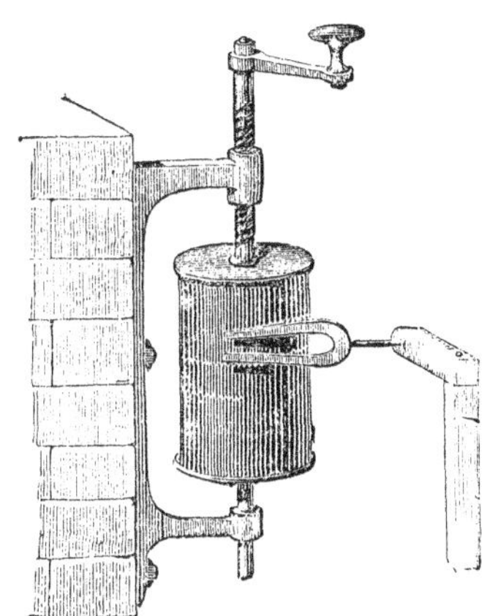
Fig. 72.

In Fig. 72 we are able to see how the tuning-fork may be made to write. Fixed to one of its prongs is a bit of pointed copper wire or a pen-nib. Observing the direction in which it vibrates, this is brought up to the cylinder in such a way that its oscillations are parallel to the axis. Before any vibration takes place the point will trace upon the revolving cylinder a fine straight line, but as soon as the vibration begins the line grows tremulous, and each sinuous curve corresponds to an oscillation of the sonorous body. The same experiment may be also tried with a plate or membrane on to which is fixed a perpendicular point of some kind—a horse-hair or hog's bristle, or a bit of tinsel. Fig. 73 represents different curves obtained in one or other of these ways.

Leon Scott had a very ingenious idea for visibly tracing the vibrations of the voice, or any other sound transmitted by the air, with a membrane arranged after this manner. This is the principle of the instrument that Kœnig called the *phonautograph:* A membrane furnished with a flexible point is stretched over the end of a kind of ear-trumpet; it

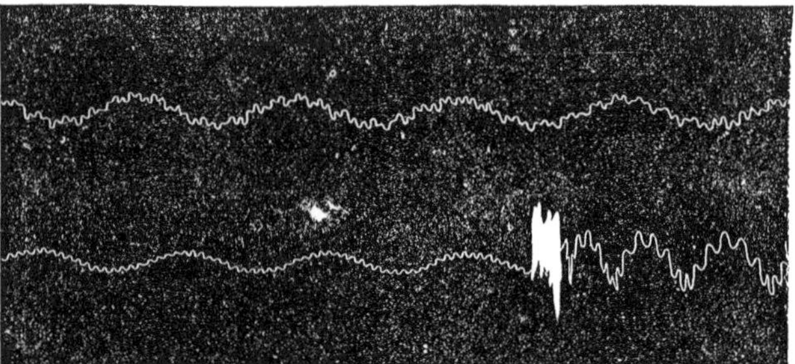

Fig. 73.

resounds loudly when a note is sounded at the other end the apparatus by the voice or an organ-pipe, and the poin will write its vibrations on a turning roller. Kœnig wrote musical air of seven notes by this means; but it is hard likely that anything more complicated could be written, f the tracings are, in general, not very intelligible.

CHAPTER IX.

PITCH OF SOUNDS.

Measure of Notes — Chladni — Mersenne — Pythagoras — Sonometer — Savart's Rattle — Sirens — Limits of Sound — Extent of the Scale of Musical Sounds — Limits of the Human Voice.

WE have seen that the origin of sound must be sought in the vibrations of elastic bodies. These vibrations are essentially *isochronous*—that is to say, the same phase continually returns at the end of the same interval, and each oscillation lasts exactly the same time as the preceding. It will be easy now to define the pitch of sounds, or that which distinguishes a low tone from a sharp one, as the duration of their vibrations, or the number of vibrations accomplished during a certain time.

Sounds of the same pitch, whatever they proceed from, correspond in the number of their vibrations. Two notes produced with different instruments are always in unison, if they have the same number of vibrations. When a note is higher than another it is because of its more rapid vibrations. Therefore, to appreciate the exact pitch of a note, the number of variations it executes in a second must be counted. One of the simplest means of ascertaining this is as follows:—The sonorous body is furnished with a point wherewith to write upon and a rotating cylinder covered with blackened paper, and is then sounded. By the side is placed a registering chronometer, which marks each second on the same cylinder. The number of zigzags, counted between the two marks, gives the pitch of the note. If the tone of a tuning-fork were known exactly beforehand, it would answer instead of the chronometer; as writing side by side with the sonorous body, whose vibrations are to be counted, each bend of its course represents a known fraction of time. Suppose, for example, that the tuning-fork makes 100 vibrations in a second, and that side by side with 50 of its oscillations 220 are found in the parallel tracing: from this we conclude that the tracing will give 440 vibrations in the time which the tuning-fork takes to accomplish 100—that is to say, in a second (Fig. 74).

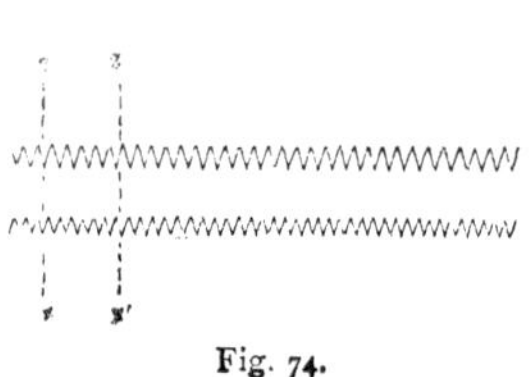

Fig. 74.

Chladni discovered a clever plan for ascertaining the number of vibrations, by starting from oscillations slow enough to be discernible, but too slow to act upon the ear. He took a metallic bar, long and thin enough to give only four oscillations a second —easy to count, watch in hand. According to the theory, a bar of half the length must give sixteen vibrations; a bar one-fourth the length, sixty-four, and so on. Continually shortening the bar, in the given proportion, we enter at last the region of sonorous vibrations. But all this only holds good in theory; in practice it is full of error.

Mersenne measured the pitch of notes by the length of the string required to produce them. He had noticed that when two strings of different lengths, but otherwise identical, were made to vibrate, the number of the vibrations was always in the inverse ratio to their length. Thus a chord of fifteen feet, stretched by a weight of seven pounds, gave ten vibrations a second; these were too slow to be heard, but by shortening the chord to one-twentieth of its length Mersenne obtained a sound twenty times sharper, or 200 vibrations a second, which he took as the starting-point for his measurements.

The sonometer or monochord (Fig. 75) acts on this principle. Its use is to ascertain the pitch of a note. On a wooden box are fixed two bridges *a*, *b*, over which a string

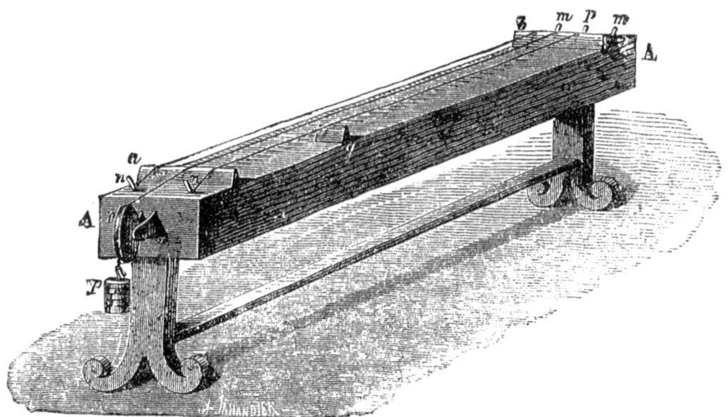

Fig. 75.

or wire is passed. One end is firmly attached to a pin; the other, being carried over the pulley *p*, is stretched by a weight. Between the two bridges is a divided scale, along which passes a movable bridge *g*, which is used to reduce the length of the string, if so required, till it is in unison with the given note; then the scale will show to a fraction the length of the chord, and a very simple calculation gives the corresponding note, provided only the note of the entire string is first known. This is settled by comparison with a tuning-fork, and we shall presently see how that is fixed.

By the sonometer it has been demonstrated that the half of the string gives the upper octave of the note rendered by the whole length of the string; that if its length is reduced to two-thirds the sound mounts to the fifth; that taking three-fourths we obtain the fourth, &c. When the entire length gives doh, the three-fourths will give fa, the two-thirds sol, the half the octave doh, and so forth. These relations existing between the length of the strings and the notes of the scale were not unknown to the Pythagoreans; and we may interpret them by saying that the octave, the fifth, and the fourth are intervals characterised by the relations of $\frac{2}{1}$, $\frac{3}{2}$, $\frac{4}{3}$ of the number of vibrations. Hence a note is the upper octave of another when it makes twice as many vibrations in the same time; also two notes have the interval of a fifth when three vibrations of the one correspond to two of the other; and they form a fourth when one makes four vibrations while the other makes three.

The sonometer also gives us a true idea of the value of the anecdote told by so many authors. One day, it is said, Pythagoras passed a forge where four blacksmiths were at work, and to his surprise he heard that the four hammers beating in measured time on the anvil gave the intervals of the fourth, the fifth, and the octave. He had them weighed, and found that their relative weights were as the numbers 1, $\frac{4}{3}$, $\frac{3}{2}$, 2. On his return home the great philosopher resolved to test this result by another experiment. He took a chord, and strained it successively by four weights equal to those of the hammers. The four notes produced under these circumstances gave the intervls of the fourth, the fifth, and the octave. Unfortunately, however, the notes of a chord do not vary in true proportion to the weight attached; to obtain the octave, for instance, we must not only double but quadruple the amount of tension. With the four weights of the hammers Pythagoras would never have been able to get these intervals from his string. Again, it would be very difficult to find hammers giving notes proportioned to their weight; the circumstance is merely a coincidence. Finally, it must be allowed that in a forge we do not hear the blow of the hammer on the bar so much as that of the bar on the anvil.

Modern scientific men have applied another principle to the measurement of the number of vibrations. It consists in producing sounds by a succession of periodical impulses given by a wheel, whose turns are registered by a mechanical contrivance. This idea was first put in practice by Stancari. He took a wheel three feet in diameter, and fixed on its outer circle 200 iron points. Thus prepared, the wheel was set on a horizontal axle, and turned with great rapidity. The points whistled through the air, and the pitch of the sound thus obtained was in proportion to the rapidity of its rotation.

About the year 1830, Savart found another method of illustrating this by a kind of huge rattle. The sounds were produced by causing the teeth of a rotating wheel to strike in quick succession against a flexible metal plate. The wheel was set in motion by a leather band passing over a large fly-wheel, which was turned by a handle. A registering apparatus fixed to the axle marked the number of turns made in a given time. Multiplying this by the number of teeth, we have the number of the vibrations executed by the edge of the plate, and consequently the pitch of the note sounded. The difficulty of turning the wheel with uniform velocity, and the bad quality of the sounds emitted by this cumbrous apparatus, have long ago brought it into disfavour.

Savart thought to supersede the siren of Cagniard de Latour by his great rattle. The plan of the siren is as follows:—A disc, perforated with holes placed in concentric circles, is rotated in such a way that a current of air is directed against a point of the perforated circle; the air passes whenever it meets a hole, and is interrupted when it strikes upon the plate. If the disc turns ten times in a second, and the holes are twelve in number, the jet of air will pass 120 times in a second, and this will also be the number of vibrations of the sound produced. This arrangement, first invented by Seebeck, is valuable in many researches. By it, for instance, it is proved that sound can only be engendered by puffs or impulses, succeeding one another at regular intervals, for the holes must be equi-distant on the disc if we want to obtain a sound corresponding to their number. Holes irregularly distributed only give a noise of high and low sounds.

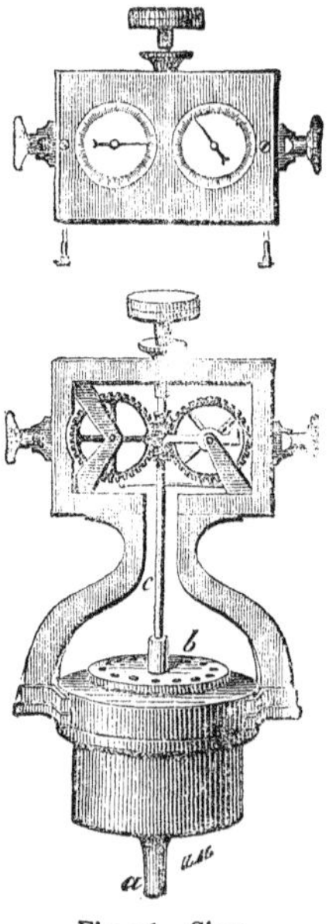

Fig. 76.—Siren of Cagniard de Latour.

The disc may be turned by a fly-wheel, or by a kind of clockwork, which also registers the number of turns. The improved siren of Cagniard de Latour (Fig. 76) was worked by the very current of air which caused the sound. The wind coming from a bellows (Fig. 77) enters through *a*, into a brass cylinder, closed at the top by a perforated disc. On this disc rests another, perforated in the same manner, which turns on an axis *c*; when the holes coincide the air passes, but it is periodically intercepted. The perforations are made obliquely through the two discs, in such a way that when the holes meet they are at right angles one with the other. Thus the current urged from below suddenly changes its direction in passing from the lower to the upper hole, and gives an impulse to the movable disc sufficient to turn it. The velocity of the rotation increases continually, and the note rises in pitch, so that if the pressure of the bellows were kept up the shrillness would become almost unbearable. It is true that the speed and the pitch may be adjusted by arranging the pressure, but it is very rarely that a perfectly regular note is obtained from the siren. When the note in unison with the one to be measured is reached, the pressure is maintained constant, while the index is consulted for the number of turns. This reckoner, shown uncovered in the figure, is set in motion by an endless screw, fixed upon the axis of the moving disc *c*; this works into two toothed wheels, which, by indices on the dials, mark respectively the hundreds, tens, and units. If, at the end of five minutes, the first dial points to 66, and the other to 30, the number of turns accomplished would be 6,630; supposing, then, the disc has twenty holes, that would give 132,600 puffs of the sonorous current in five minutes, or 300 seconds, or 442 a second; from which we conclude that the note obtained corresponds to 442 double vibrations

The siren can sing under water, and therefore gained its name. Plunged in any liquid, it can be made to sing by forcing a powerful jet of the same through the aperture

Thus water, oil, and mercury will sing. The sounds are distinguished by a peculiar quality, but the notes are the same as in the air.

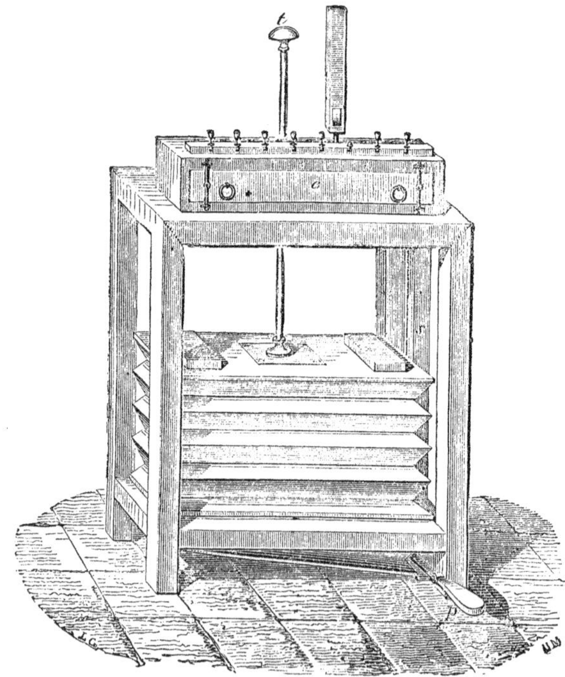

Fig. 77.—The Bellows.

We must plainly confess that the tone of the siren is not so pleasant to the ear as its name would lead one to suppose; these shrill and piercing sounds would scarcely set us dreaming of the Siren's songs which Homer says allured travellers by their wondrous spell, and if we stop our ears, it is certainly not for fear of being bewitched.

To produce the current of air requisite for working these instruments, an apparatus is used (Fig. 77) composed of a double pair of bellows, acted upon by a pedal *p*, a rod *t*, and an air compartment *c*, perforated with a certain number of holes. By these holes the siren, or the tubes which are to be sounded, receive the wind. They can be opened and shut at pleasure by pressing different buttons.

A natural question arises here as to the limit of audible sound. What are the very lowest and the highest notes appreciable by the ear?

In 1700, Sauveur pronounced the lowest sound to be that produced in a pipe of forty feet, corresponding to twenty-five vibrations per second.

The deepest bass-pipe yet constructed by organ-builders is thirty-two feet in length. It should give the doh$_{-2}$, corresponding to thirty-two simple vibrations. On the other hand they make very short pipes, which should give 10,000 vibrations, or more. But is it proved that these sounds actually exist?

The lowest notes of the octave of sixteen feet, the doh of sixty-five, and the re of seventy-three vibrations, are heard only as a kind of rumbling, in which the most practised ear can scarcely distinguish the musical pitch; and the pipes that produce these notes can only be tuned by indirect means. On the piano, where they constitute the lower extremity of the key-board, their musical character is very undecided; and orchestral music but rarely descends below the mi of the double-bass, which has eighty-two vibrations. In these regions the ear already begins to apprehend the vibrations of the air as separate shocks. This sensation becomes more distinct as we advance to the octave of thirty-two feet, and as we approach the doh of thirty-two vibrations we no longer hear a sound, properly speaking; that which strikes the ear is only a series of disconnected explosions. Many people, nevertheless, imagine that they have heard the notes of this octave ; but this is because the organ-pipes produce, simultaneously with their fundamental note, other higher notes of which we shall speak presently; a pipe of thirty-two feet causes the notes belonging to a higher octave to resound slightly, and this in all probability deceives the listener.

The same illusion is doubtless present in the conclusions Savart has drawn from his experiments on the limits of hearing. He arranged a bar of iron to turn round a horizontal axis in such a manner, that at each half revolution it should pass through a chink hollowed in a plank. At the moment of its entrance the bar forced the air like a piston, producing a sort of explosion, and if the wheel turned fast enough a deep sound was heard, accompanied by a loud rumbling. Seven or eight revolutions a second still gave an audible sound, wherefore Savart concluded that the deepest note distinguishable by the ear might be fixed at seven or eight double, or fourteen to sixteen simple vibrations. But Despretz has without difficulty shown the error of this, for by arranging two chinks instead of one for the iron bar to pass through, we do not get the octave as we ought by doubling the number of the explosions. It must then be admitted that the note of thirty-two vibrations, corresponding to sixteen rotations, has already been obtained by eight; and this is not surprising if we remember that natural sounds are almost always accompanied by higher notes, called harmonics, as we shall presently see. At the most, Savart's instrument gives a note of about thirty semi or simple vibrations a second.

Helmholtz had recourse to another plan. He used a wooden case closed at both ends, and having a small opening into which was fitted a gutta-percha tube, intended to be introduced into the auditive canal. On this sounding-board he stretched a wire, weighted in the middle by a brass coin with a hole in it; owing to this precaution the string could not give the upper octaves of its fundamental note, which was very deep.

Under these circumstances, the sound of a string which gives a medium note becomes insupportable through its strength; but that employed in these experiments, giving the re of seventy-three vibrations, produced only a faint and slightly growling noise. Coming down to si of sixty-one vibrations, Helmholtz scarcely heard anything. From these experiments he concluded that audible sounds began at about sixty semi-vibrations, and took a musical character at about eighty, in the octave already mentioned of sixteen feet. But the limits of hearing may perhaps vary in different persons, and depend in some degree on experience and on the intensity of sounds.

The higher limit of hearing is certainly not the same for every one. Many people cannot distinguish certain high notes that others hear perfectly. Savart tells us that a sound of 31,000 semi-vibrations, produced by the longitudinal vibrations of a glass cylinder, was heard by the greater part of his audience, whilst the 33,000 vibrations of a cylinder a little smaller were scarcely heard at all. With large toothed wheels he produced a very intense sound, which was not lost till the moment when it appeared to perform 48,000 vibrations per second; but it is difficult to prove in this case that the flexible plate touched all the teeth of the wheel.

Despretz thought to extend this limit by means of tuning-forks which should give 73,000 semi-vibrations. There are some miniature tuning-forks still preserved at the Sorbonne, and shown on special occasions. But how are the notes determined? M. Marloye first adjusted tuning-forks to the ear. He began by making a scale which passed from 16,000 to 32,000 vibrations, guiding himself by ear; then in the same manner he tuned a fork to an octave higher than the last, giving consequently 64,000 vibrations, and corresponding to doh_{10}; then he went to re_{10} of 73,000 vibrations. These tuning-forks can only be heard by very sensitive ears; the very shrill notes produce a painful impression, an indefinable uneasiness which lingers for some time; it is very difficult to perceive their musical relations. Till further light dawns on the subject, we do not deem these conclusions very important.

Recently Kœnig resumed these experiments. The highest notes that he could distinguish corresponded to 40,000 vibrations; but, as we have already said, the limit varies with different persons. Very high notes cease to be appreciable by many ears. Has not Wollaston told us that many people are quite incapable of hearing the sharp chirp of grasshoppers, or even the twittering of sparrows? Perhaps there are animals who can distinguish notes beyond the reach of human ears.

To resume: appreciable sounds are limited to a range of from about 60 to 40,000 semi-vibrations per second, which range may be sometimes passed for ears of exceptional power and delicacy. The undulations of the ether produced by light and heat are infinitely more rapid. Heat begins at 65,000,000 vibrations, visible colours range from 400 to 900 trillions, and chemical rays attain as much as a quadrillion. Heat is not produced simply by the vibrations of the fluid ether; it is certain that ponderous bodies themselves vibrate when they are heated; therefore we must admit that molecules can accomplish vibrations of wondrous rapidity. But what becomes of those vibrations which are too rapid to be audible, and too slow to be felt as heat? Have we senses that can appreciate them, organs that can be affected by them? May we seek in these unclassified vibrations the explanation of galvanism and electricity, which everything leads us to suppose a form of motion? Who can tell?

It will not be uninteresting to mention here the compass of the notes given by the commonest musical instruments. First stands the organ, the grandest and richest of all, which occupies the whole field of audible vibrations—nearly ten octaves. The piano has almost seven octaves, comprising all the notes from la$_{-2}$ to doh$_{7}$, or from 54 to 8,400 vibrations.

The sounds of the violin properly extend from 400 to 6,000, along four octaves, but much higher sounds can be drawn from this instrument. The violoncello, or violone, is confined to a scale of between 80 and 350 vibrations; but the octo-basso of M. Villaume embraced vibrations as low as 64. The cornet, trombone, and other brass instruments give very varied sounds. The highest note used in the orchestra is probably the re_7, which corresponds to 9,400 vibrations.

We may take as the extreme limits of the human voice the fa_{-1} of 87, and the doh_6 of 4,200 vibrations—

CHAPTER X.

THE NOTES.

Relation of the Notes—Scale—Names of the Notes—Hymn of St. John—Musical Notation—Major and Minor Keys—The Waves of the Tempered Scale—Galin and Chevé — Choir and Concert Pitch —Natural Tuning-fork—M. Lissajous' Method.

MUSIC is not so much concerned with the absolute pitch of notes, as with their relation one to another, or the intervals between them. The pleasure we derive from the combination of certain sounds depends on this relation. When two notes are in the mutual relation of two simple whole numbers, they form a concord or harmony; discords are produced by complex relations. In this sense we may say that music is a matter of numbers.

Pythagoras was aware that a string divided into two unequal sections would give two perfectly harmonious sounds, when the lengths of the two sections hold a simple relation to one another, expressible by whole numbers. The relation 1 : 2 corresponds to the octave; the relation 2 : 3 to the fifth; 3 : 4 to the fourth, and so on. Most probably the Greek philosopher had learnt this law from the Egyptian priests, which is equivalent to saying that it was known in the earliest times.

Harmonious intervals, therefore, are based on the relations of the pitch of the notes. Take, for example, the fifth doh, sol. The ear tells us that this harmony may be found between very high as well as between very low notes, not at all depending on the absolute number of vibrations. Measurements show that any two notes having this interval hold always the mutual proportion of 3 : 2, and consequently this interval is always caught by the ear when two notes are as 3 : 2. From this it is easy to see that the more nearly this relation is consummated, the purer and sweeter will be the harmony; and therefore this interval is called a *true fifth*. We shall presently see that it is seldom realised in all its purity.

The simple intervals adopted by musicians are characterised by the following relations:—

Octave	1 : 2
Fifth	2 : 3
Fourth	3 : 4
Major third	4 : 5
Minor third	5 : 6
Major sixth	3 : 5
Minor sixth	5 : 8

A note is said to be the upper octave of another when it makes twice as many vibrations in a given time, and *vice versâ*. The successive octaves of a note are distinguished by figures placed below or in a bracket, thus: doh_2 means the upper octave of doh (we never write doh_1); doh_3 is the upper octave of doh_2, or the double of doh, &c. Descending to the lower octaves we write them thus: doh_{-1} is the lower octave of doh, doh_{-2} the double octave, and so on.

It is easy to see that two, three, or four notes which harmonise when taken two and two, will still accord when united altogether. The two chords of three notes most pleasant to the ear are the perfect major chord, characterised by the numbers 4, 5, 6, and the perfect minor chord, represented by the fractions $\frac{1}{6}$, $\frac{1}{5}$, $\frac{1}{4}$. They both contain a fifth, a major third, and a minor third, the only difference being that in the major chord the major third is the lower, while in the minor it is the upper interval. To realise the different harmonies a musical *scale* has been adopted, composed of seven degrees (the octave of the first note making an eighth), which may be expressed by the following syllables :—

Doh, re, mi, fa, sol, la, si, doh ;

the relation amongst them being as the numbers—

24, 27, 30, 32, 36, 40, 45, 48.

The first scale is followed by another, and so on, each being formed by raising all the notes of the preceding scale one octave. We have already described how the successive octaves are written. The relations which the different notes of the scale bear to the first, constitute their musical intervals, and are expressed by the following numbers :—

Doh—doh	unison	1 : 1
Doh—re	second	8 : 9
Doh—mi	third	4 : 5
Doh—fa	fourth	3 : 4
Doh—sol	fifth	2 : 3
Doh—la	sixth	3 : 5
Doh—si	seventh	8 : 15
Doh—doh_2	octave	1 : 2
Doh—re_2	ninth	4 : 9
Doh—mi_2	tenth	2 : 5
Doh—fa_2	eleventh	3 : 8
Doh—sol_2	twelfth	1 : 3
* *	* *	*
Doh—doh_3	double octave	1 : 4
* *	* *	*
Doh—mi_3	seventeenth	1 : 5
&c.	&c.	&c.

The names of the intervals simply recall the position of the notes in the scale. The twelfth, the double octave, and the seventeenth make perfect harmonies, which fact presupposes the simplicity of the relations which characterise them; it is needless to particularise them further, since they are but the counterparts of the fifth, the octave, and the third.

Associating the notes of the scale by twos, we do not always obtain a harmony. A suitable choice must be made. But even discords are important in music. The interval from doh to re, called a major tone; the interval from re to mi, called a minor tone; the intervals mi—fa and si—doh_2, known as diatonic semitones, are very characteristic discords.

The scale just explained does not suppose any knowledge of the absolute pitch of the notes; it merely depends upon the relationship they bear one to another. The first note may be anything; but its value once determined, that of all the other notes is fixed also. This may be noticed in the exercises of solfeggio, which consists in singing the notes of the scale on the syllables doh, re, mi, fa, sol, la, si. The sound represented by doh may be chosen arbitrarily; but by this choice the pitch of all the notes is decided. If, for example, the doh has 240 vibrations, the re must have 270, the mi 300, the fa 320, and so on.

The names of the first six notes were introduced in 1026, by Guido l'Aretino, or Guy of Arezzo; they are the beginnings of words taken from the hymn of John the Baptist:—

"*Ut** queant laxis *re*sonare fibris
*Mi*ra gestorum *fa*muli tuorum,
*Sol*ve polluti *la*bii reatum,
Sancte Ioannes."

The air to which this hymn is now sung at St. Jean is not exactly the same as the ancient air, in which the six syllables chosen by L'Aretino really fall upon the notes they name. That air, found in a MS. in the library of the Chapter of Sens, has been copied in old style, as follows:—

* *Ut* is the first syllable in French, but is replaced by *doh* in English.

HYMN OF ST. JOHN

Ancient Melody.

The seventh syllable, si, was not added till 1684, by Lemaire. In Italy they soon substituted doh in place of ut, as being a more vocal syllable. The names proposed by Guy did not come quickly into general use, for in the time of Jean de Muris, in the fourteenth century, they still used the syllables pro, to, no, do, tu, a, in Paris; but at last they were accepted pretty generally, excepting in England and Germany, where they kept for the notes the names of the letters C, D, E, F, G, A, B, or H.

Here is the history of the letter designation. Since the time of Gregory the Great, perhaps even before the sixteenth century, a series of scales of fixed notes corresponding to the limits of the voice and to the sounds of the principal instruments had been used. They were called after the first seven letters of the alphabet, in this way :—

A, B, C, D, E, F, G, a, b, c, d, e, f, g, aa, bb, cc, &c.

At a later date, a note having been added below, it was designated by the Gamma, or Greek G, whence comes the common name of the scale, Gamut.

Guido l'Aretino substituted for these letters points set upon parallel lines (*les portées*), to each of which a letter served as key. The key fixed the value of the line; thus, when F was written upon the beginning of a line, all points

L

placed upon this line represented the note F. Afterwards they enlarged these points, and determined to place them in the intermediate spaces, and multiplied both lines and spaces, as it was found necessary.

The signs of the notes only served at first to mark the difference of intonation, without respect to the duration. Jean de Muris, or Mœurs, invented square figures to distinguish the relative value or duration of the notes. This was about the year 1338, and in 1502 the invention was perfected by Octavio Petrucci, who discovered a way of printing music with movable type. The longest note according to the old notation was called a Long, ; the next in duration was a Breve, or . Of these, the latter is occasionally found in church music, the former but seldom. The moderns have gradually confined themselves almost entirely to the following, which, since the fifteenth century, have been indicated by the accompanying signs :—

Semibreve. Minim. Crotchet. Quaver. Semiquaver. Demisemiquaver.

We also find occasionally— Hemidemisemiquaver.

A long is equal in duration to two breves, a breve to two semibreves, a semibreve to two minims, a minim to two crotchets, and so on. These notes may be replaced by equivalent rests :—

Long Rest. Breve Rest. Semibreve Rest. Minim Rest. Crotchet Rest. Quaver Rest Semiquaver Rest. Demisemiquaver Rest.

To fix the absolute duration of a note, a metronome is employed.

The letter G has become the key of sol, ; the letter F, the key of fa, ; the letter C, the key of doh, , &c.

The syllables doh, re, mi, fa, sol, la did not originally designate any fixed notes, but only the degrees of a scale; they represent the hexachord of Guido l'Aretino. They used to be written underneath the letters which marked the fixed scales, beginning with C, F, or G.

C	D	E	F	G	A	B	c	d	e	f	. . .
doh	re	mi	fa	sol	la	..	..	..	..	..	
..	..	..	doh	re	mi	fa	sol	la	..	..	
..	..	..	..	doh	re	mi	fa	sol	la	..	
							doh	re	mi	fa	

The same fixed note might then occupy different places in the movable scale, and this was sometimes found to be incompatible with the preservation of the intervals adopted for the notes doh, re, mi, fa, sol, la. This led to different plans for harmonising, and there was a great confusion in the musical system. The necessity was soon felt of altering some of the fixed notes, when the movable scale was so transposed, that the intervals of the fixed corresponding notes did not realise the intervals first intended by the notes doh, re, mi, fa, sol, la. Thus, when doh was written below F, and fa below B, the interval from F to B should have been a fourth; but as it was in reality greater, it was lessened by lowering B a semitone. This note then became B flat, while it remained B natural in the scale beginning with C. This double part it had to play was indicated by writing the B in different ways, and it is also the origin of the signs (♭) flat, and (♮) natural.*

It was only after a thousand changes and attempts that the modern musical system took form. The principal rule which directs it is this: Whatever note be fixed upon for

* This is more clearly shown by the French words *bémol* and *bécarré*.

beginning the scale, the other notes must all follow in the intervals already decided on. To provide for this necessity the sounds are altered, either by raising them a semitone, which is called sharpening, and this is expressed in the notation by the sign ♯; or by lowering them a semitone, which is called flattening, and is expressed by the sign ♭. For the value of this semitone the ratio $\frac{24}{25}$ is used, which is less than $\frac{15}{16}$, the value of the interval from mi to fa.*

The words doh, re, mi, fa, sol, la, si are now used for the principal fixed notes of the piano and other instruments, and following the sign ♭ or ♯ they become changed notes, in such instruments as the organ and pianoforte. In vocal music and fidicinal instruments, the ratios of the diatonic scale are preserved in every key. The scales always bear the name of their first note or tone. All the major scales are modelled on the scale of doh, formed by the set of natural notes—

Doh, re, mi, fa, sol, la, si, doh.

The intervals are reproduced with tolerable exactitude, owing to the alterations applied to certain notes. The scale of sol is composed of the notes—

Sol, la, si, doh, re, mi, fa♯, sol;

the scale of fa, of the notes—

Fa, sol, la, si♭, doh, re, mi, fa;

and so forth. These scales belong to the major key. There have been many other scales used in music which, from having the third minor, have given rise to minor scales, as, *e.g.*—

La, si, doh, re, mi, fa, sol, la.

The chief difference between the two scales lies in the

* The semitone is nearer fa than mi.

introduction of the minor third, la—doh (5 : 6), in place of the major third, doh—mi (4 : 5); they are each characterised by a perfect harmony formed with the third and the fifth of the tonic.

Perfect major chord . . .	doh, mi, sol.
Perfect minor chord . . .	la, doh, mi, *or* doh, mi♭, sol.

The minor scale is still further varied by raising the seventh, and sometimes also the sixth note of the scale a semitone, for certain harmonic reasons.

It would singularly complicate the construction of all instruments with fixed sounds, if it were attempted to make them realise the scales in their theoretical purity. It was necessary to make a compromise, and this was done in the tempered or adjusted scale. The ear will tolerate a slight deviation from perfect harmony, and this allows a simplification of the scale in instruments with fixed notes, by employing only one sound for two notes nearly alike, from the inverse alteration of two neighbouring notes. Thus doh♯ and re♭ have but one pipe or string for both, &c. &c. In this way it is managed on a keyed instrument to interpolate five black keys with the seven white of each octave, thus forming the chromatic scale, which is composed of twelve equal semitones, adapting themselves to all the exigencies of the musical system. It follows that we are thereby led to alter more or less sensibly the natural notes represented by the white keys, and so to modify all musical intervals.

The adjusted semitones may be approximately rendered by the relation $\frac{18}{17}$; and an adjusted whole tone scarcely differs from a major tone $\frac{9}{8}$. The fifth and the fourth are only falsified to an inappreciable extent by the adjustment, but the thirds are so much so that they are painful to an ear educated to pure harmony, which is wonderfully more exquisite. Some authors of the last century gave the name of "wolves" to these lost intervals, where the discords seemed to meet and growl.

A natural voice, guided only by instinct, always gives true intervals; and violinists whose ears have not been spoilt by the orchestra will play true thirds and sixths much more delightful than the adjusted intervals. Unfortunately the free-toned instruments, which play in the orchestra with tempered or adjusted instruments, are forced to follow their lead and acknowledge the false intervals; and thus those violinists who have all their lives been forced to play falsely in the orchestra, become accustomed to the change of tone. Under the overwhelming influence of the orchestra the accuracy of the voice also suffers. Singers end by adapting themselves to the adjusted notes, and lose the power of singing a simple air with that true intonation which constitutes its charm. Still, if a singer have true ear and taste, Nature reasserts her rights as soon as she is relieved from the requirements of the accompaniment.

The inconveniences arising from the equal adjustment have given rise to numberless attempts to return to natural harmony, even in instrumental music. Erard's harp with a double movement; Poole's enharmonic organ, and that of Gen. Perronet Thompson; the harmonium devised by Helmholtz—all give the different scales without the aid of adjusting or tempering. The vocal systems adopted in France by Galin and Chevé, and in England by the numerous Tonic Sol-fa Associations, hold to the natural scales in their purity. The English societies employ the syllables doh, re, mi, fa, sol, la, ti, doh, and reduce them in writing to the letters d, r, m, f, s, l, t, d. Galin and Chevé employed the figures 1, 2, 3, 4, 5, 6, 7, for this purpose, the successive octaves being indicated by points placed over or under the figures. It is only needful to give the absolute pitch of the first note, or tonic, for all the other notes to be determined. This plan is believed to give greater facilities for reading music than the old notation, and has had many advocates. Rousseau recommended it most highly.

"Music," says J. J. Rousseau, "has shared the fate of all arts which are only brought to perfection slowly. The inventors of notes thought merely of the state of the art in their own day, not looking on to the future; and, therefore, the nearer the art draws to perfection, the more defective are their signs found to be. As it advances, new rules are established to obviate present inconveniences; in multiplying the signs, the difficulties also are multiplied; and what with additions and alterations, they have formed out of a simple principle a most cumbrous and ill-arranged system. Musicians, it is true, do not admit this. Custom is everything. Music is not for them the science of sounds; it is but a science of crotchets and quavers and minims. As soon as these are lost to sight, they think that music is done with. Besides, why should that be made easy for others which they have acquired with such difficulty? The musician is not the one to be consulted on this subject, but a man who understands music, and has reflected on the art."

When a piece is to be played by several performers, it is necessary for the instruments to agree; therefore, in the orchestra they are tuned by means of a tuning-fork, whose note remains constant. Formerly, the pitch used to be given to an orchestra by a kind of whistle, furnished with a graduated piston, whereby the pipe could be lengthened or shortened at will, so as to draw different fixed sounds from it. There was the *choir-pitch* for the plain song and for secular music, the *chapel-pitch*, and the orchestra or *concert-pitch*. The latter was never fixed: they raised or lowered it, according to the compass of the voices. The chapel-pitch, on the contrary, was fixed, at least in France, and generally higher than concert-pitch. As for the choir-pitch, which agreed with the organ, it is hard to say whether it was higher or lower than the chapel-pitch, for authors contradict one another on this point; it would seem that after all they only set the organ to chapel-pitch.

Since the science has been possessed of means for measuring the absolute pitch of notes, musicians have been able otherwise to determine the pitch of the different leading orchestras of Europe, and, very curiously, it has been discovered that it is everywhere rising rapidly. Sauveur, who appears to have first studied the question, found in 1700 that the lowest note in the harpsichord, la, made 202 vibrations; and the low doh of the harpsichord, 244 vibrations, which gave la_3 810. Other determinations of the last century vary from 820 to 850. In 1833, Henri Scheibler examined the tuning-forks of the principal theatres, and found that at the Opera they had two of 853 and 868; at the Italian and Conservatoire, others of 870 and 881 vibrations; at Berlin he found a la of 883; at Vienna they varied from 867 to 890. In 1857, M. Lissajous declared a new progression in the orchestra-pitch. Here are the results of his measurements:—

Opera of Paris	896
Opera of Berlin	897
Theatre of San Carlo, Naples . . .	890
Theatre della Scala, Milan . . .	903
Italian Opera, London	904
Maximum in London	910

This increasing elevation in the pitch of instruments is proved by the ancient organs found in some basilicas. What is the reason that musicians and authors have made this change? It is supposed that most instruments show greatest brilliancy in their high notes, and therefore the makers have by little and little heightened the pitch. Singers generally follow the same inclination, to the detriment of their voices. But we must not go too far in attributing the ruin of so many fine voices to this solely; it would be fairer to seek the cause, as M. Berlioz does, in the tendency of modern composers to write higher parts for vocal music than the ancient composers. Whatever the height of the pitch may be, it is easy for the composer to keep within reasonable limits.

It is none the less true, however, that the progressive variation must at last trouble the musicians, and it is very important to return to a natural and absolutely settled pitch. Sauveur insisted on the necessity of this so long ago as 1700. He first proposed for this purpose the sound which makes 200 vibrations per second. Finding subsequently that his calculation was erroneous, he modified his views, and so proposed to take a doh of 512 vibrations for his starting-point. This number is one of the series 1, 2, 4, 8, &c., whose terms may be regarded as the successive octaves of unity. Chladni afterwards adopted the same doh of 512 vibrations, corresponding to the natural la, 853, and this was generally employed by scientific men. However, as the pitch of the orchestras continued to rise, the German philosophers meeting at Stuttgard in 1834 decided on choosing a normal la more in harmony with the custom of musicians, and they fixed definitively on the la of 880 vibrations; this is the German la, most useful for numerical calculations. Unhappily, this congress could not reach the rest of the world, and the pitch still mounted in a very disorderly manner. Then it was that the decree of February 16, 1859, fixed an official diapason for France. This pitch gives the normal la with 870 vibrations; it scarcely differs from the German, yet it is much less useful for calculations.

Here follow the numbers of the simple vibrations of the adjusted scale based upon la_3 (French style), and of the natural scale beginning with the same doh. The octaves are obtained by doubling, or by dividing by two.

Notes.	Adjusted Scale.	Natural Scale.	Natural Ratio or Relation.
Doh	... 517.3	... 517.3	... 24
Re	... 580.7	... 582.0	... 27
Mi	... 651.8	... 646.6	... 30
Fa	... 690.5	... 689.7	... 32
Sol	... 775.1	... 776.0	... 36
La	... 870.0	... 862.2	... 40
Si	... 976.5	... 970.0	... 45
Doh	... 1034.6	... 1034.6	... 48

The middle octave of the piano is represented by the following notes :—

Henceforth, in France all musical instruments will be tuned by a tuning-fork set to the official standard of the Conservatoire. Concord is thus ensured, and there is no more to fear from the tendency of orchestras to raise the pitch.

The piano, violin, and other instruments are generally

tuned by ear. One string is set to the note of the tuning-fork, and the others are regulated by the musical intervals, chiefly by octaves and fifths. According to Weber's experiments, a very fine ear can appreciate a difference of a thousandth part, or one vibration in a thousand, but that is the limit. The study of beats, however (a phenomenon which we shall soon notice), leads us much further. It is by this means that organs are tuned. When extreme precision is required we have recourse to a later method invented by M. Lissajous the principle of which will now be explained.

A prismatic rod can vibrate transversely, so that its free end describes a right line. If a steel bead be fastened at this end, the continuance of the luminous impressions will appear as a line of brilliancy. The eye has the power of preserving the most fugitive impressions for about the fifteenth part of a second. If then the luminous point run its course in less time than $\frac{1}{15}$ of a second, the whole track will appear illuminated. Thus a burning stick or piece of charcoal swung round in the air will make a fiery circle. When the section of the rod is rectangular, it can be made to vibrate either in its thickness or its breadth. In either case the bead will draw a line of light, but in the former the route will be perpendicular to that which it takes in the latter. But we may agitate the rod in yet another way by striking it obliquely. It is then moved simultaneously in two directions crossing at right angles. Will it decide to follow one impulse rather than the other? The rod takes a middle course between the two roads, and follows first one impulse and then the other, changing momentarily. The little bead takes a tortuous road, and its luminous track allows us to follow the rod in its rapid evolutions.

The number of straight vibrations depends on the direction in which the rod vibrates. When the section of the rod is square, its thickness and breadth being equal, the number of vibrations will evidently be the same in both directions. In this case the little bead will describe an ellipse, which may either pass into a circle or flatten to a straight line. Calculation proves this. The line may be understood *a priori* by supposing that the rod moves diagonally from its position of repose, always making little equal steps forwards and to the right, forwards and to the right; then, in returning, backwards and to the left, backwards and to the left, as shown in Fig. 78. To explain the ellipses, it would be necessary to enter upon some rather abstruse propositions.

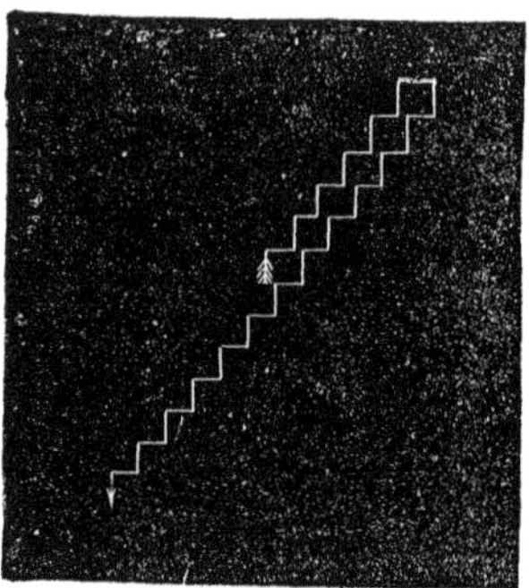

Fig. 78.—Vibration of a Square Rod.

When the two dimensions of the rod are as 1 : 2, the corresponding numbers of vibrations will evidently be in the relation of the octave; if the measurements are as 2 : 3, the vibrations will be the fifth, &c. The bead and the reflected ray then will describe the curves given later on in Figs. 84 and 85. It may therefore be said that these figures characterise the musical intervals.

Wheatstone's kaleidophone (Fig. 79) is on this principle. This is an apparatus composed of several metal rods, to the end of which are fixed light glass beads, silvered within. When illuminated by the sun or the light of a lamp, the bright spots will describe curves as shown in the figures, while the rods vibrate. Wheatstone made this known in 1827, and the kaleidophone is now found very frequently in the studios of scientific men. Let us, then, examine some other conclusions drawn from the same principle. Imagine an upright mirror fixed at the end of a horizontal bar, which can be made to vibrate alternately vertically and horizontally (Fig. 80). On this mirror we throw a luminous ray, by placing it before a lamp covered with a shade, from

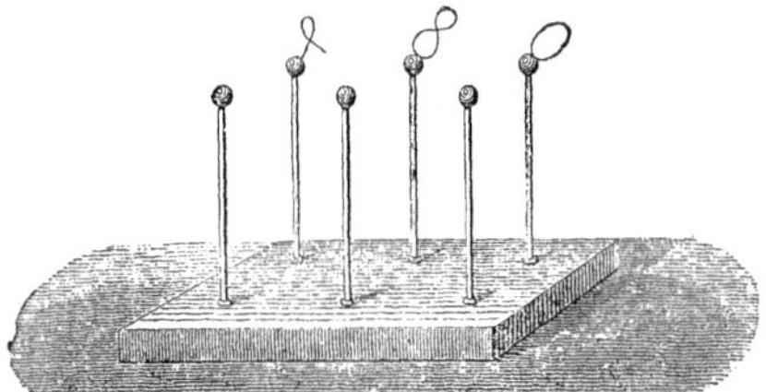

Fig. 79.—The Kaleidophone.

which a single ray escapes by a small hole pierced for the purpose. While the mirror remains motionless the reflected ray will form upon the wall a simple point of light; looking straight into the glass for the image of the lamp, we see the tiny light shining like a fixed star. Now, if the bar be struck so as to oscillate, the reflected ray shares the movement of the mirror, and the image on the wall is displaced. As, at first, the bar only vibrates in a vertical plane, we see upon the wall a luminous track, drawn straight down; and looking into the vibrating mirror, we see there also a perpendicular line. If, on the contrary, a horizontal motion be given to the bar. the reflected line will be horizontal too.

Finally, if the bar be made to vibrate obliquely, we shall see both upon the wall and mirror the fanciful curves of the kaleidophone. It will even be sufficient to hold before the mirror a metal button, a pin's head, or any small bright object. Its reflection will form a luminous curve as soon as the bar is set in motion. The form of the curves will always depend on the rate of the vibrations executed by the bar in a straight line, if it oscillates first in a vertical, and then in a horizontal plane.

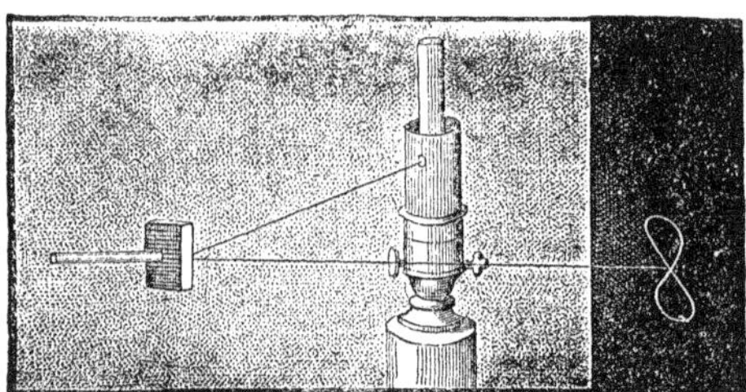

Fig. 80.

The same curve may be obtained by a double reflection on two mirrors, each vibrating in a different plane (Fig. 81). They are placed opposite one another, so that a ray of light reflected by the first will fall upon the second, which throws it back in its turn against the wall. If, then, one only of the mirrors be made to vibrate, the brilliant point upon the wall will change into a luminous line, drawn in the direction of the vibrations, because the reflected ray shares the motion of the reflecting surface. But if the first mirror be made to vibrate horizontally, and the second vertically, the reflected ray will receive from the first a horizontal movement, to which is added a vertical movement by the second reflection; the two movements combine, as in the kaleidophone, to give birth to the different curves already

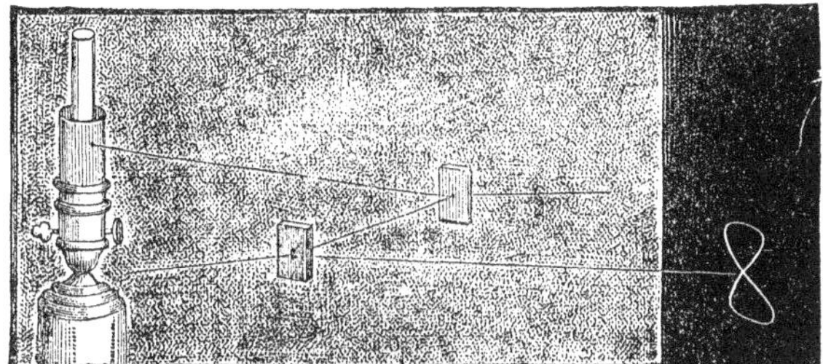

Fig. 81.

described. They may be seen either by looking directly into the second mirror, or by receiving the image of the luminous point upon a screen of any kind. Greater clearness and

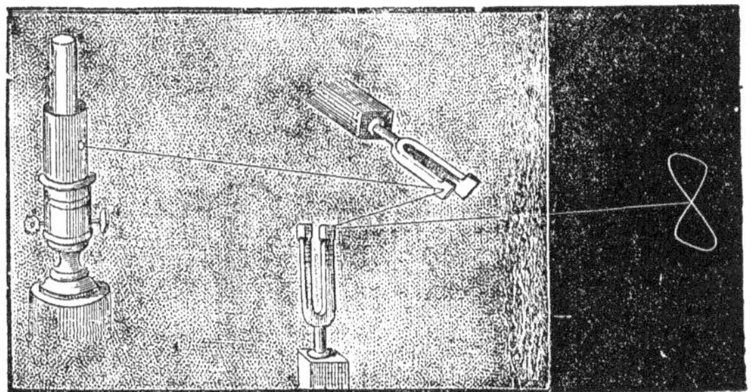

Fig. 82.—The Optical Method of M. Lissajous.

brilliancy is given to this experiment by passing the luminous rays through a lens. A simple inspection of the curves will shew the ratio of the respective numbers of the vibrations made by the two mirrors. A straight line or an ellipse indicates unison, the figure 8 the octave, and so on.

Instead of fixing the two mirrors to horizontal and

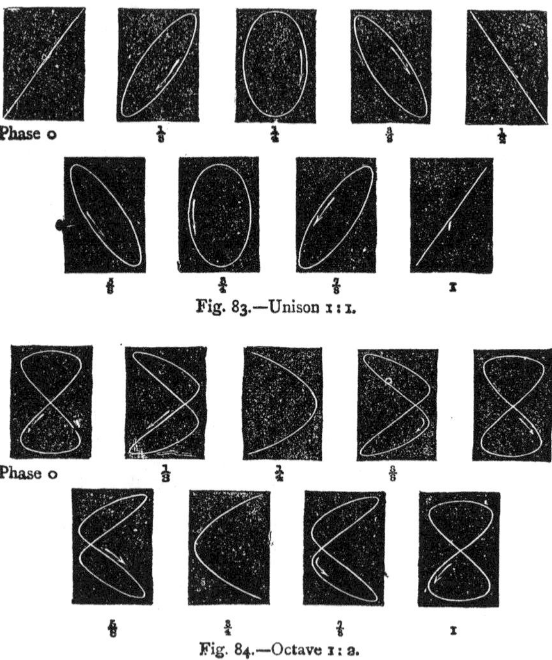

Fig. 83.—Unison 1 : 1.

Fig. 84.—Octave 1 : 2.

vertical rods, they may be fastened against the branches of two tuning-forks, placed at right angles, one horizontally and the other vertically, as in Fig. 82. The first gives to the reflected ray a horizontal movement, the second imparts to it a vertical impulse, and thus are obtained curves which reveal at once, by their aspect, the musical relation of the two forks. Herein consists the optical method for comparing sonorous vibrations, made known by M. Lissajous in 1855. It enables us to ascertain the musical interval of

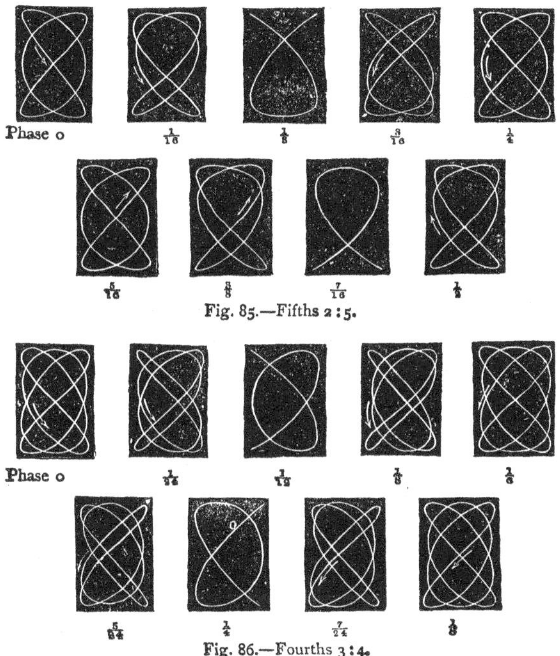

Fig. 85.—Fifths 2 : 5.

Fig. 86.—Fourths 3 : 4.

two vibrating bodies, with a certainty unknown before this beautiful discovery.

It may be asked why the same ratio should produce different figures. This is due to the difference of phase. If one of the two mirrors be slightly behind the other in first beginning to vibrate, this delay (which is called difference of phase, or simply phase) modifies the appearance of the figure resulting from the combination of the two movements. Thus, when two tuning-forks in perfect unison begin and end their course together (when there is no phase), the trajectory of the luminous image is a right line; in any other case it is an ellipse or a circle. Under each figure will be found written the difference of phase as a fraction of the entire vibration.

When the vibrations of two tuning-forks are in the ratio of two whole numbers, the optical figure drawn at the beginning of their movement will continue unchanged as to form, but will diminish slightly in size as the vibrations die away. In this case, only one of the curves which characterise the musical interval in question will be seen. But if there be the slightest discordance between the two tuning-forks, the figure does not remain steady, but changes gradually, so as to pass through a complete cycle of the different curves which correspond to the same interval. This is because the delay (or phase) continually increases, and the figure consequently changes in the same way. The more decided the discord, the more rapid the changes. So it happens that the ellipse which characterises unison will pass into an oblique ellipse crossing the line, then narrowing into the form of a straight line, it passes on to a reversed obliquity. This variation of the figures betrays the slightest discord immediately, and also helps to an appreciation of its value.

By this means the tuning-forks are tested in the Conservatoire; once corrected by the standard there, they are stamped and fully recognised. But the fork to be tested has no mirror attached; its own polished surface serves the purpose.

M. Lissajous has added to his beautiful inventions that of the "vibration microscope." The object-glass is held by one branch of a tuning-fork placed at right angles with the tube. When the fork vibrates the object-glass oscillates before the tube, and the objects upon the field of the microscope seem to oscillate in the same direction. If, then, one of the objects itself vibrates in a different direction, the real and the apparent vibrations blend, and the curve thereby formed will show the number of vibrations of the body under consideration.

CHAPTER XI.

TIMBRE OR QUALITY OF SOUND.

Form of Waves—Simple and Complex Sounds—Harmonics—Timbre of Voices and Musical Instruments—Musical Sounds—Vowels.

WE have seen that the pitch of a note depends on the rapidity with which the vibrations succeed one another. Is that the only difference which can exist between sounds? Evidently not; for we never confuse sounds having a different origin, even when they are in unison; they are distinguished by what has been called *timbre*. The sounds of the cornet, for instance, do not resemble those of the harp, nor does the violin sound like the organ. The same note, even, has a different character according as it is sung on *a* or *o*; whence it follows that the vowels only represent the changing timbre of the human voice. We may even classify the differences in the timbre of musical instruments by determining which vowels they seem most to resemble.

What, then, causes the timbre? How can the same note produce such different impressions? These questions have long occupied philosophers, and it is but latterly that they have been satisfactorily answered, owing to the researches of Helmholtz.

It had always been supposed, and with reason, that the timbre must have some connection with the particular form of the vibrations of the sonorous body. Their number simply determined the pitch; no other possible difference remained than that which might be presented by each vibration taken separately. Such a difference was easily discoverable in liquid waves, which may be pointed, crested, or flattened, while still keeping the same rate of vibration. A puff of wind ruffling the surface of the water causes numberless little ripples, which change the form of the waves without hastening or retarding their motion. But what is the form of a fixed vibration (like that of a chord), where each of the points of the vibrating body simply rises and falls, and therefore always remains in the same straight line? Nothing is simpler. Just as a man might go from one place to another in a thousand different ways during a quarter of an hour, loitering the first five minutes, then running a little way, and again dawdling at the end of the journey, so a vibrating particle may change in more than one manner during the hundredth part of a second which it takes to run its course. It can go first slowly, then very fast, and again slacken its speed; and it may do this two or three times along its route. The revolving mirror enables us to record the alterations of velocity which take place during one simple oscillation. A sheet of smoked paper, which is moved rapidly under the vibrating point, will show in visible tracery all the irregularities of the oscillating motion; by looking at the curve so obtained, it may be known at once how many times during each oscillation the *andante* alternates with the *presto*. The revolving mirror reflects a bead fixed at the end of a horizontal bar, in a series of different perspectives, giving the appearance of a luminous ribbon; if, then, the bar vibrate perpendicularly to this ribbon, the bead rises and falls, and the shining band changes into a chain of serpentine folds.

The curve is exactly analogous to that shown by the graphic tracery.

When the particular nature of a periodical motion is known beforehand, the curves may be traced without having

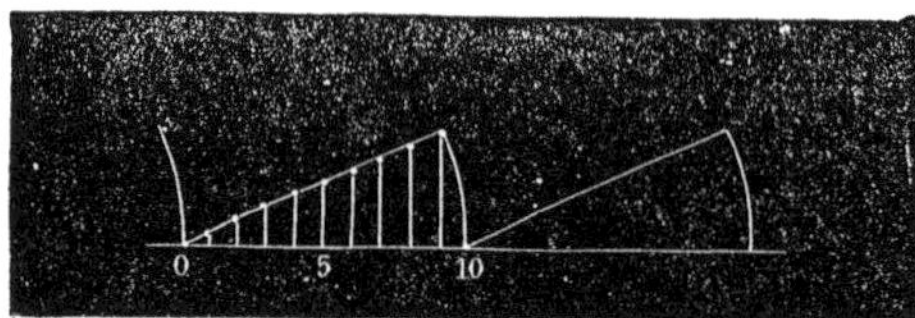

Fig. 87.

been seen. On a horizontal line the successive seconds must be marked; at each division an upright line is raised to the height where the vibrating body should be found at this moment; the extremities of these lines give the curve of the vibration. Thus Fig. 87 represents the periodic

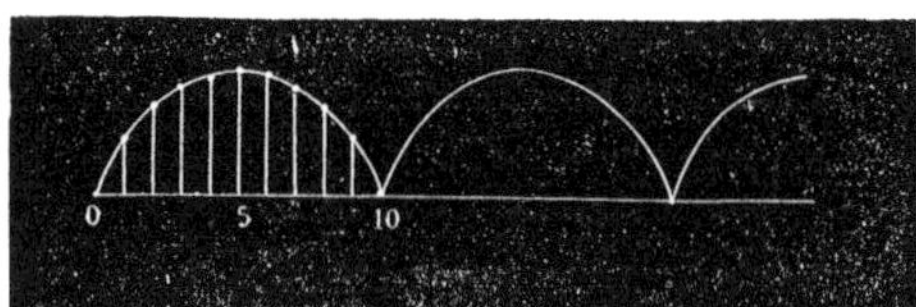

Fig. 88.

motion of a hammer worked by a hydraulic wheel: first it rises slowly, then suddenly falls; at the first point it is quite low, up to the ninth it lazily rises, between the ninth and tenth it comes down with a sudden fall. The motion of the bow-string of an archer is just the same. Fig. 88 shows in like manner the course of an india rubber ball which rebounds vertically after having touched the ground. A revolving mirro would show it describing this curve, which is formed of successive arches.

The simplest or most regular periodical movement is that of the pendulum. It is represented by a curve having the sinuous form of Fig. 89. Thus a pendulum ending in a point will trace its oscillations on a sheet of paper slipped underneath it. The straight line indicates the direction in which the paper is drawn; the oscillations are perpendicular

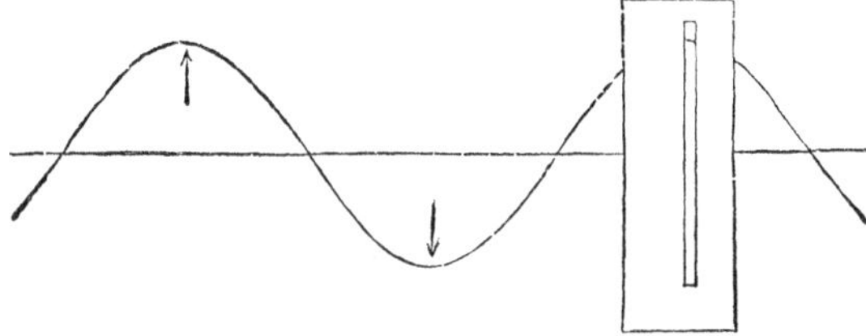

Fig. 89.

to this line, as pointed out by the arrows. It is easy, by the aid of this curve, to reproduce the remarkable movement of the well-known simple pendulum. Take a card, and after cutting a slit in it with a penknife, hold it against the curve in such a position that the slit shall be vertical; then move it slowly from right to left. You will never see more than one point of the curve, and it will seem to oscillate in the slit just like a pendulum.

The mathematical law of pendular motion may be in some degree explained by illustration. Let us imagine a luminous point—a small lantern, for instance—fastened to the edge of a vertical wheel revolving with a uniform velocity (Fig. 90). Placing yourself opposite this, you will see the light describe a perfect circle. The appearance would be very different viewed sideways. Take a somewhat distant position, where you see only the edge of the wheel, and the light will seem to travel up and down exactly in a perpendicular line, only it will have the appearance of going much faster in the centre of the line than at the top or bottom. At these two points, indeed, it will seem to stop for a moment before turning. Now this apparent movement will

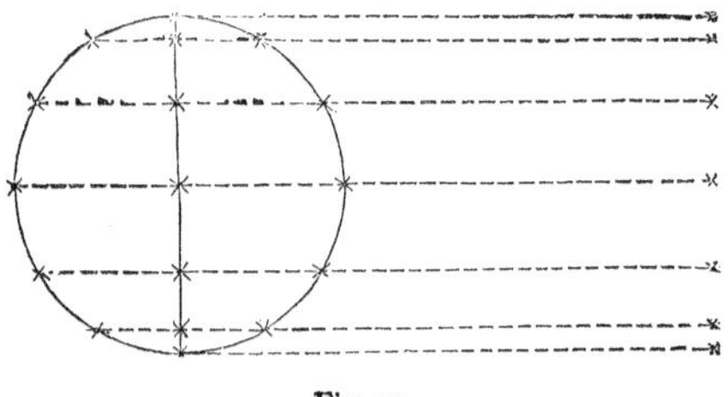

Fig. 90.

be the exact imitation of a pendular movement, which would make the luminous point swing the length of the vertical diameter of the wheel.

A "pendular vibration" is any periodical movement of the same character as that of the pendulum, the velocity being zero at the two extremities, and increasing towards the middle, where it reaches its maximum. A simple sound is produced by a pendular vibration. The motion of the branches of a common tuning-fork approach this type of vibration; it gives a note very nearly simple, and so also does the flute.

All simple sounds are exceedingly sweet, and seem softer

than they really are. Their timbre has something mournful, recalling the timbre of the vowel combination *ou;* this is quite independent of the material of the sonorous body. We shall soon see what is necessary to produce a simple sound; it is a *rara avis* of nature, seldom if ever met with.

The sounds we find in nature are complex, that is to say, they are composed of several simple sounds differing in height. Each body forms a little orchestra to itself when it vibrates freely. The lowest sound gives the pitch, the others accompany it. This it is that gives the timbre or tone. A rich, full timbre is like a nest of harmonious sounds, whose warblings please us, we know not why.

It had long been known that many bodies give fainter sounds at the same time with the fundamental one; and these were called harmonics; but no one understood the part they played, nor was it suspected that they are the principal, if not the only, cause of the *tone* distinguishing different instruments, and that the numberless vibratory curves are explained by their intervention.

Sauveur gave the name of *harmonics* of a fundamental sound to those sounds which make 2, 3, 4, 5 vibrations, whilst the other makes only one; together they form the natural series of 1, 2, 3, 4, 5. The first harmonic is the octave of the fundamental sound, and the second is its twelfth, or the octave of the fifth; then follow the double octave; the seventeenth, or the double octave of the third; the nineteenth, or double octave of the fifth, &c.

In order to indicate the ratio of the height of the harmonics by their designations, the fundamental sound has been included with them, as the harmonic 1; the octave will be the harmonic 2; the twelfth, the harmonic 3, &c. Taking doh$_2$ for the fundamental sound, we have the following series:—

Notwithstanding their names, these notes do not invariably form harmonious chords. The first six, however, do so; 7 and 11, approximately represented by la♯ and fa♯, do not even belong to the musical scale; they are discordant notes, and so is 9, the re. When these notes are perceived in a compound sound they mar its beauty, giving it somewhat of a harsh or jarring tone.

In 1700, Sauveur thus notices the phenomenon of harmonics or "overtones:"—

"On striking a harp-string," says he, "besides the fundamental sound, a number more may be heard at the same time by a delicate and educated ear, sharper than that of the entire chord, produced by some portions of the string which, freeing themselves in some way from the general vibration, take one of their own. These complex vibrations may be explained by the example of a slack-rope, such as dancers use; for while the rope-dancer gives the rope a violent swing, he may with his two hands give two different impulses to the two halves."

"Each half, each third, each quarter of a string has its own special vibrations, while the general vibration of the whole string is going on. It is the same with a bell when it is very good and tuneful."

After enumerating the successive harmonics which accompany the fundamental sound of a string, he adds: "It would appear, then, that whenever Nature makes for herself, as we may say, a musical system, she employs sounds of this kind; and yet they have been hitherto unknown to the theory of musicians. When they were heard, they were treated as irregular and of no consequence, the musicians thinking thereby to prevent a breach in the imperfect and limited system then in vogue."

Twenty-five years later, Rameau used these ideas as the base of a new musical system.

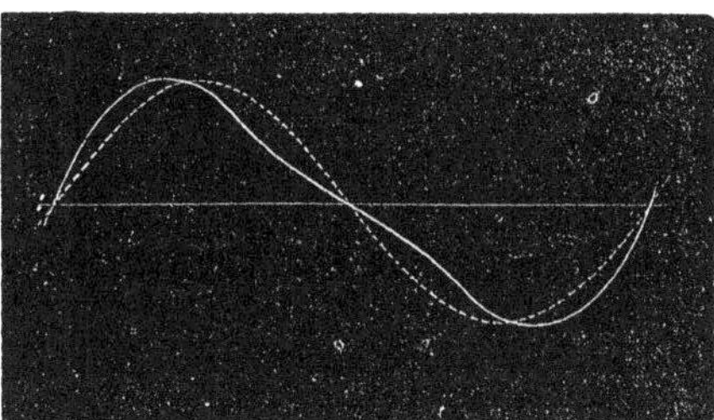

Fig. 91.—Fundamental Sound and Octave.

The fundamental sound and its harmonics, taken singly, are simple sounds with pendular vibration. Their intermixture constitutes a complex sound, whose vibrations take a form more or less complicated. Each of these compounded vibrations is composed—1st, of one vibration of the fundamental sound; 2ndly, of two vibrations of the octave; 3rdly, of three vibrations of the twelfth; 4thly, of four vibrations of the double octave, and so on. The general form of the curve which represents this compound vibration, is determined by the fundamental sound; but the harmonics make its contour shrink and swell by their vibrations. In Fig. 91, the dotted line represents the curve of the fundamental sound, and the white line, the curve resulting from the addition of the octave. It is a curve of this species which characterises the timbre or quality of a compound sound; it changes form according to the relative intensity of the harmonics; but the number of the great curves or periods is always the same, and for this reason, the pitch of the mixed sound is that of the fundamental note.

Inversely, a periodical vibration, of whatever form, may always be separated into a series of simple harmonic vibrations of pendular form. In other words, all complex sound of a definite pitch may be resolved into an harmonic series of simple sounds, beginning with the fundamental, which has the same pitch as the complex sound. This is a theorem of Fourier's, and one of the most interesting ever drawn from analysis; but we cannot make more than a passing reference to it. From it we conclude, that if quality depend on the form of vibrations, this form in its turn depends upon harmonics, so that in reality quality is given by the superposition of simple sounds. This is no mathematical fiction, no subtle definition devoid of reality; experience confirms these deductions in the most striking manner.

In order thoroughly to understand a compound movement, let us refer once more to the undulations of a liquid surface. Suppose the water be agitated by two stones, in two different places; there are then two centres of commotion, whence two systems of circular and concentric rings spread out, till they meet and interpenetrate; but the eye can still follow their separate circles. It is beautiful to watch this kind of motion at the sea-side. The waves as they come in, easily distinguished by their foaming crests, break in a regular succession, and thrown back in different ways, according to the form of the coast-line, they intermingle, crossing obliquely in all directions. A steam-boat leaves behind her in the water two divergent breaks of dancing waves ; a bird plunging after a fish will make a succession of tiny circular waves, which work their way across the general commotion. It is rarely that an attentive observer fails to follow the different partial movements which give a special form and direction to each.

In the same way the ear can perfectly distinguish the different sonorous movements transmitted to it simultaneously by the air. Let us transport ourselves in thought to the midst of a ball-room, at the moment when the orchestra bursts forth with a merry dance. What a mixture of sounds, which can nevertheless be disentangled more or less ! The strings of the bass violin and the mouths of men give out sonorous waves twelve or fourteen feet long ; the rosy lips of women give shorter and more rapid undulations ; the silken rustle of dresses, and the noise of footsteps, produce small tempests of tiny crowded waves ; and all these mingle without losing their identity, for the ear can still distinguish their different origin. The auditory canal, however, which receives all these impressions at once, is but a speck in comparison with the mass of air in the room where all these vibratory motions are going on. The ear cannot follow the sonorous waves throughout their course, as the eye observes the motions in a sheet of water.

If a stone be thrown into water already agitated by undulations of a certain extent, little concentric circles will be seen spreading over the undulated surface, just the same as in quiet waters. At the moment when the little circular ring coincides with the crest of one of the great waves, the height of this wave is suddenly augmented by the height of the little one; so too its tiny depression, added for a moment to the depression already existing between the large waves, will hollow it yet a little more. On the contrary, when a depression meets with an elevation, the principal effect will be weakened. Thus the addition of smaller waves to greater simply increases the height of the hollow; and if we can imagine the little wavelets raised out of our vision for a moment, we shall see nothing but the large waves, slightly modified in their outline.

The separation of elementary notes, found associated in any noise whatever, may nevertheless be effected by the ear with the aid of the resonant globes already described. We have seen that these globes each reinforce a particular note of which they are constituted guardians; they respond to, echo, and draw it, so to say, out of the general tumult. With a set of these globes, each made for a special note, it will be easy to single out the notes from any medley, however slight their existing force. Thus also it is proved that the harmonics of musical sounds, far from being a fanciful illusion, a merely subjective phenomenon, have a true existence. With a little practice, they may be caught by the ear alone.

Once accustomed to listen, the ear listens almost unconsciously. Thus when a drum is heard at some little distance, a low dull sound is first noticed, which is caused by the air imprisoned in the hollow; then a succession of sharp notes, clearer and more defined, produced by the stretched parchment or head; other harsh sounds are due to the jarring of the strings on the lower parchment; and finally there is a metallic ring, coming from the sides of the cylinder.

The human voice is very rich in harmonics, taking very complex timbres. With the sympathetic resonant balls, sixteen harmonics or overtones can be reckoned in a bass voice singing *a* or *e*, on a very low note. Rameau was not unaware of this phenomenon, and many musicians have noticed it since. Seiler tells us how, in listening during sleepless nights to the voice of the watchmen telling the hours in Leipsig, he often seemed to hear first the twelfth, and then the note itself. M. Garcia says that, listening to his own voice in the silence of night upon a bridge, he has been able to distinguish the octave and the twelfth of the note he gave. We seldom notice the existence of these parasite notes in the sound of the voice, because we do not look for them ; but we may easily convince ourselves of the fact in this way. Ask a singer to sing the vowel *o* on the mi♭ in the bass, then gently strike the si♭ of the middle octave on the piano, so as to fix attention on this note. You will continue to hear the si♭ after the finger has left the piano and the string has ceased to vibrate. This is because the si♭, resounding in the mi♭ of the voice, will replace the sound of the string. If you wish to try the sol of the following octave, or the seventeenth of mi♭, in this way, it will be better to take the vowel *a*.

Let us mention here that the notes from mi_6 to sol_6, belonging to the last octave of the piano, are always heightened in tone by a peculiar resonance they excite in the auditory canal ; thus acquiring a fictitious intensity, which gives a piercing character to the sounds they accompany as overtones. To a sensitive ear it is actually painful. We know that even dogs are very sensitive to this kind of impression ; a high mi on the violin will make them howl. This irritability of the ear in regard to very high notes, renders it particularly sensitive to those disagreeable dissonances that always strike us in choirs, especially when the voices are at all forced. Above the lower notes we really hear a crowd of little screaming notes, accompanying the harmony like an orchestra of castanets and cymbals.

Fine strings also abound in overtones. Helmholtz has counted as many as eighteen. The harmonics 7, 9, 11, 13, 14, 17, 18, are more or less discordant; if they had more intensity they would produce a most unpleasant effect. Happily the ear only catches the first upper notes, which agree with the fundamental note, and even these can only be seized by close attention.

These facts seem to show that *all* sonorous vibration, having a peculiar timbre, is reduced by the ear to simple sounds which form an harmonious series. This conclusion may seem at first sight too absolute, and contrary to our senses, since we are not accustomed to take note of the existence of several notes in a musical sound. At most, musicians only distinguish in a chord the notes that form it, but that are produced separately. The difficulty seems to augment when the chord is formed with compound intervals, such as the twelfth, repeat of the fifth, and the seventeenth, triplique of the third (as Sauveur calls it). Kœnig made a pretty experiment in this way. On the sounding-board of an enormous tuning-fork he arranged a whole orchestra of small ones, which gave amongst them the first four or five harmonics of their leader. Then with a vigorous stroke he set the great patriarch in vibration, and afterwards all his attendants: the air was filled with a deep harmonious sound, very full, but seeming to the unpractised ear a single note, the voices of the sharper forks not being heard. He then suddenly stifled the deepest by placing his hand upon it, and the others were heard immediately—clearly separating themselves, as soon as the deep tone which had sustained and bound them together was subdued.

Thus, in ordinary circumstances, the ear seems unable to accomplish the dissection necessary for reducing the timbre to its constituent parts. But this is a mistake. It is only necessary to understand the words we use. Indeed, we must here distinguish between *perception* or *sensation*, which is complex, and the *impression* received by the mind, which is simple. The ear really perceives several notes when fa is given by the violin, but the whole of these notes only recall to our mind a fa having a peculiar timbre; we have no particular reason for analysing our impression further. The hearing apparatus dissects the complex sounds that strike it, but the separate elements are reunited in the nervous impression made on the mind. Physiology gives us many instances of similar illusions. Thus we take for simple colours, tints which the prism divides into numberless tints. The theory of binocular vision shows how during our whole lives we see all objects double, and nevertheless it needs a strong effort of attention to be convinced of it. Few people know that in the retina there is a little blind spot, the *punctum cœcum*, and that consequently in one direction we cannot see at all. This blank is so large that there would be room in it for seven lunar images in a row, and at a distance of a few feet a human face would be lost in it, yet it is not inconvenient. When Mariotte illustrated the fact by experiments in the court of Charles II., he was greatly amused at the astonishment occasioned among his illustrious audience. There are some well-authenticated instances of people who have only discovered by chance that they had lost the sight of one eye years ago. Such is our indifference to a phenomenon always present with us. We do not notice the complexity of a sound any more than the double image of an object that we look at with both eyes; yet it is this very duplicity that gives the effect of relief, as shown by the stereoscope. Timbre is the *relief* of sounds.

We manage to distinguish the sounds of different instruments, or the voices of different people; and in these cases there are many things to help us besides the timbre—those little noises which precede and follow the emission of the sound, its duration and power, its intermissions and variations. But the ear must be educated to the task of dissecting the timbre, in order to be *conscious* of its complexity.

Helmholtz has corroborated these deductions by composing different artificial timbres with the notes they were supposed to contain. Here is an experiment that any one can easily try: Raise the hammers of a piano so as to have all the wires at liberty, then sing loudly the vowel *a* upon any note you choose, standing near the instrument. The resounding of the strings exactly reproduces the *a*. The resemblance is much less complete when the hammers are not all lifted from the strings; because the vowel *a* is characterised by a peculiar timbre, depending on certain sharp notes; the strings corresponding to these notes vibrate through sympathy, and their intervention gives to the echo of the voice the timbre it had in singing the *a*. In the same way the timbre of the clarionet, the cornet, and so on, may be imitated.

The height of a musical sound, then, is always that of the dominant note in this harmonic medley, and this is generally the lowest of all. But the presence of the upper notes is not without its influence on our judgment of a complex sound—it sharpens it, slightly raising the musical scale. For this reason even practised musicians sometimes mistake an octave in comparing notes of different timbre.

We have already said that the ear does not depend solely upon timbre in discerning the origin of sounds, but is guided by certain accessory noises. In many cases these characteristic noises are only heard at the first moment, or as the sound dies away.

The preparation for the emission of a sound is almost as important as its timbre. With the human voice the noises preceding the emission of the vowels are so very distinct, that they are called after the explosive consonants, *b*, *p*, *d*, *t*, *g*, *k*. They give to the vowel following a peculiar character quite apart from its timbre.

In any loud note given by a brass instrument, we can distinguish between the hautbois, clarionet, &c., without regard to the timbre. Then, too, the greater or lesser rapidity with which the fundamental sound and its harmonics die away, constitutes a sensible difference between catgut strings and wires, even when they are equally struck. The vibrations of the first being unsustained, their sound is somewhat poor and dry; while, the vibrations of the metal wires enduring much longer, their sound is fuller, though less penetrating.

In other cases the sound is accompanied by noises throughout. Thus, in wind instruments there is a sort of whistling, caused by the action of the air on the edge of the opening. The scraping of the bow is always heard more or less with the violin. Noises of this kind are expressed by the letters *f*, *v*, *s*, *j*, *z*, *l*, *r*.

The vowels, too, are constantly accompanied by little noises, that help us to guess them even when they are whispered. These sounds are heard more in speaking than in singing, for in singing the timbre or the musical part of the vowel is most dwelt upon, and this is heard to a much greater distance. This is why consonants are not heard so far away as vowels, and why a distant voice may be mistaken for a cornet. The consonants *n* and *m*, however, by their mode of formation, have somewhat of the nature of vowels, and the accessory noises play a very subordinate part. If you stand at the foot of a hill and listen to voices speaking some way up, you will scarcely catch any words except those formed with *m* or *n*.

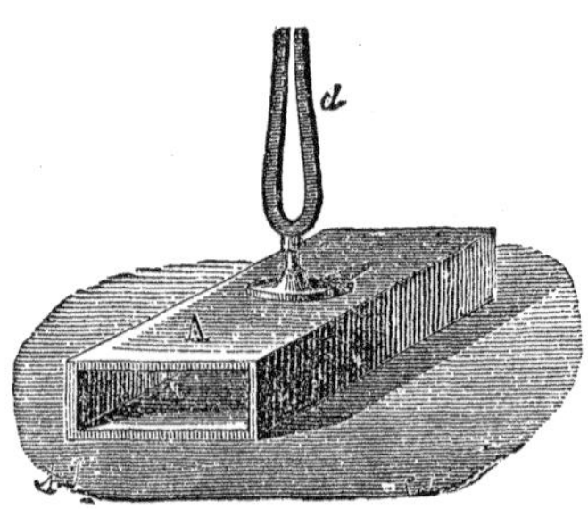

Fig. 92.—Mounted Tuning-fork.

A few further remarks may be made about different timbres. In the first place, we can obtain simple sounds by strengthening the fundamental sound of a tuning-fork by a resounding box (A, Fig. 92), whose upper notes do not harmonise with those of the fork. The timbre of simple sounds is sweet and subdued, not brilliant enough for music.

Sounds accompanied by upper notes that are not harmonious, are not included in our definition of musical sound: we can only use them in music when the upper notes die away so quickly that we may forget them, and notice only the principal note. In this category we place rods, discs, tuning-forks, bells, parchment skins, &c. Tuning-forks have very high upper notes, heard at the moment of striking the metal. The first is at an interval of a twelfth from the fundamental sound. The ear always separates these sharp quickly passing notes from the principal notes, and has no tendency to blend them with it, as it does the harmonious elements of a musical sound.

The sound of common bells can hardly be ranked as a musical sound: but it appears that a skilful founder is able to make the first upper notes of a bell harmonious, and then the timbre is tolerably good. This explains the pleasant effect of chimes. There are eight at Amsterdam, one of which numbers forty-two bells, and has a compass of three octaves and a half (between doh_2 and fa_5). The most celebrated is that of Ghent. Paris is going to have one at Saint Germain l'Auxerrois.

The fundamental sound of bells is lowered by an increase of weight or diameter. The largest bell in the world is that cast at Moscow, in 1736. Its weight is about 193 tons. Unfortunately it was cracked before ever it was rung. Still, there is one at Moscow, weighing 63 tons, that dates from 1307. The great bells of our cathedrals seldom weigh more than 10 tons. That of Notre Dame de Paris, founded in 1680, weighs nearly 13 tons.

Franklin's harmonica is composed of a number of glass bells, which are sounded by rubbing round the edges with damp fingers. The effect is rather irritating to the nerves, the sound being too penetrating, because of the prevalence of harmonic overtones.

Instruments that are played upon by striking, such as timbrels, tambourines, castanets, triangles, and cymbals, are classed together with bells and tuning-forks. They have discordant upper notes. The tam-tam or gong of the Chinese is a circular disc with a raised edge, made of well-tempered and hammered bronze. It is struck with quick light taps from the rim to the centre, and wonderful effects are got from the multiplied sounds, which gather and seem to burst out with great violence. It is as if a struggle were going on in the metal of sounds which make frantic efforts to escape from their prison. The sheet iron with which the sound of thunder is imitated in theatres, produces effects somewhat similar.

The skins of the drum and tambourine do not give true musical sounds, but the resistance of the frame or body stifles the higher notes considerably. All these noisy instruments are employed chiefly to mark the time, and they are in high favour among savages. There is not a nation on the earth that has not invented a drum of some kind to beat a measure, and animate the dancers. Amongst the Esquimaux, the Patagonians and Hottentots, and the New Zealanders, they are to be found. An earthen pot or bit of hollowed wood, or a calabash, with an ass's or crocodile's skin, form the materials of these rough resounding boxes. The tambourine and the castanets, which Southern nations use so gracefully, are of very ancient origin. The *crotalon* of the priestesses of Bacchus (Fig. 94) was nothing more.

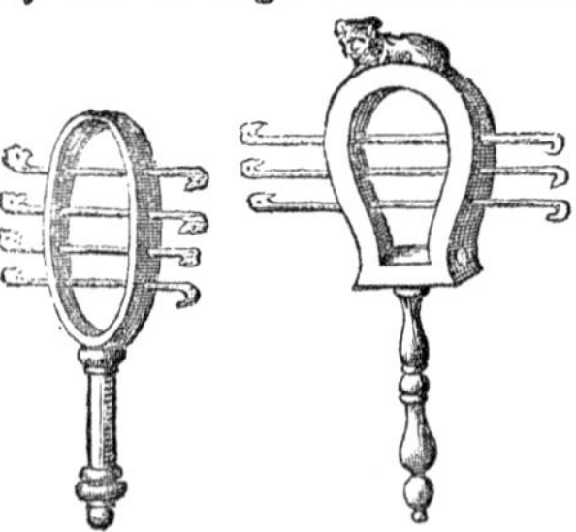

Fig. 93.—Sistra of the Ancients.

Strings and pipes are pre-eminent as the true source of musical instruments. Their timbre is harmonious. A homogeneous string vibrating completely gives, besides its fundamental sound, the series 2, 3, 4, as harmonics; but it may be made to vibrate in such a way as only to give one

FIG. 94.—PRIESTESS OF BACCHUS (from a Bas-relief).

of its harmonics, dividing itself into different segments, separated by nodes.

The quality or timbre varies, according as they are played upon by pulling, as in the harp; by striking with a hammer, as in the piano; by drawing a bow across, as in the violin; or by the wind blowing over them, as in the Æolian harp.

In the construction of pianos, the experience of two centuries led to the foundation of a number of rules, which are now justified by theory. Thus the hammers of the middle strings have been made to strike them at the seventh or the ninth of their length, because the best quality or timbre was thus obtained. Theory shows that by this arrangement the harmonics 7 and 9, the first which will not harmonise with the fundamental sound, are suppressed. The time during which the hammer remains in contact with the string also influences the timbre.

Strings of cat-gut have very little persistency of sound, though their harmonics are very high; so the disagreeable effect of these is neutralised. In the violin their timbre is slightly modified by the resonance of the instrument, whose own proper sound is generally doh_3. The first harmonics are less distinct in the violin than in the piano, but the sharp harmonics are more strongly marked.

Open pipes are much like strings, having a fundamental sound, with a timbre comprising the natural series of notes, 1, 2, 3, 4, 5; and the fundamental sound can be got rid of, and nothing left but a harmonic, by forcing the wind. In the closed pipes some of the harmonics are wanting: they give only the notes 1, 3, 5, 7.

A closed pipe has always the same fundamental sound as an open one of double the length; this may be seen by closing an open pipe midway with a slide (*p*, Fig. 95), so reducing it to a closed pipe of half its former length, when the sound will remain the same. In short, the law that explains the names of the register (or draw-stop) of an organ is this: The height of the fundamental sound is in inverse ratio to the length of the pipes. An open pipe of 16 feet gives the lower octave of the open pipe of 8 feet, but is in unison with the closed pipe of 8 feet; the open pipe of 8 feet is the lower octave of the open pipe of 4 feet, and in unison with the closed pipe of 4 feet, &c.

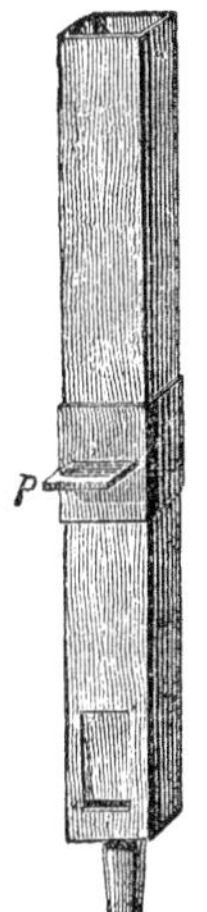

Fig. 95.

In the organ there is a pipe for every note, each one giving only its fundamental; but in other wind instruments there are many plans for getting all the notes of the scale out of the same pipe. Thus the horn is made of a very long brass pipe, curled round: its only harmonics are 8, 9, 10; but these will give the actual scale by a little modification, which is done by introducing the hand into the end. In the trombone the length of the pipe is varied by a slide; in the cornet-à-piston, by supplementary pipes. In other instruments, like the flute and clarionet, the pipe is pierced with holes, that are opened and closed by keys. The column of air in the pipe is made to vibrate in such a way as to form centres, in relation to the open holes, wherefore these openings produce the same effect as if the pipe were cut at the places where they are situated. Owing to this mechanism, the musician has in his hands a whole set of pipes of different lengths, from which he can draw the most varied sounds.

In all wind instruments one of the most important parts

is the mouth or opening. The most simple is such as we find in flutes and the generality of organ-pipes; it is represented by the whistle (Fig. 96), which is a simple mouth-piece without a pipe. The wind strikes upon the lip of the mouth with a rustling that may be considered as a medley of feeble sounds. The column of air in the pipe strengthens some of these by a sympathetic resonance, and these are the harmonics the pipe will utter. In the reed mouth-pieces the stream of air first sets in vibration a metal key, which interrupts it periodically This trembling of the key gives birth to a number of notes, among which the column of air makes its choice; but the sound is not the same as when the pipe is played with an ordinary mouth-piece. To this list belong the reed-stop pipes of organs, and the notes of the harmonium, clarionet, hautbois, bassoon, cornet, and cor anglais. Our lips act as reed mouth-pieces when playing upon such instruments as the horn, trumpet, or trombone, their position and their tension influencing one or other of the harmonics of the tube of the instrument.

Fig 96.

In the production of the voice there are vocal chords which play the same part, but their mode of action is quite different from that of the lips. They determine the height of the note for singing or speaking. In the clarionet and horn the note depends on the volume of air in the pipe; but here, on the contrary, it only depends on the tension of the vocal chords, and not at all on the volume of air which is made to resound by their action. But this resonance becomes very important from another point of view. It modifies the timbre by favouring certain sounds. This is the origin of vowels.

A vowel is nothing more than the particular timbre taken by any note, if the resonance of the mouth strengthens, amongst the harmonics of this note, that which approaches nearest to a certain fixed note. Thus, for example, the vowel *a* is produced by the resonance of si$\flat_4$. To articulate *a*, the mouth is placed in such a position as to sound si$\flat_4$; and whatever be the fundamental of the sound we emit, it is always the harmonic nearest to si$\flat_4$ which will be made prominent.

If when the mouth be opened to articulate some such vowel a number of tuning-forks of various pitch be passed before it, one will always be found to answer to it by increase of sound: its note is the one that answers to the proper volume of air contained in the mouth. In this way Helmholtz found that each vowel is characterised by one or two notes, usually the same, but sometimes modified, according to the accent in which the vowel is spoken.

It is easy to understand how this occurs. The definition of the vowels as five letters of the alphabet is altogether insufficient, as they are indeed numberless, if we take heed of all shades of pronunciation. We must at least distinguish seven principal vowel sounds which group themselves in this way:—

Therefore if a vowel be defined by its specific note, the note varies with the language in which the vowel is spoken. The notes decided on by Helmholtz for the German vowels, diff-

from those that M. Donders attributes to the same vowels pronounced in Dutch.

The vowels *a*, *o*, and *ou* have always only one single specific note, but for the others two are found; and this twofold expression is explained if we remember that the mouth in their case takes the form of a bottle, the wide part being represented by the mouth, and the narrow neck by the tongue and the lips. These two cavities vibrate separately. Here are the notes which, according to Helmholtz, answer to the vowels spoken in the accent of North Germany:—

The intensity of the partial sounds of a vowel does not, then, depend on the place they occupy in the harmonic scale, but only on their absolute pitch; and it is this which distinguishes the timbre of vowels from that of musical instruments. Take, for example, a flute: whatever note it gives, it is always the octave which resounds simultaneously. But if *a* be sung upon any note whatever, one cannot foresee what harmonic will be strengthened: sometimes it will be the octave, sometimes the twelfth or the seventeenth, or some other note of the harmonic series. Thus if *a* be sung upon the note si$\flat_3$, the octave will be given, for the specific note, of the vowel *a* is the octave of si$\flat_3$; but if the fundamental note be fa♯, the ninth harmonic la$\sharp_4$, which is the nearest to si$\flat_4$, will be heard above all. There is a slight analogy here with the violin, which always strengthens the neighbouring notes of doh$_3$, the sound belonging to the volume of air imprisoned within it.

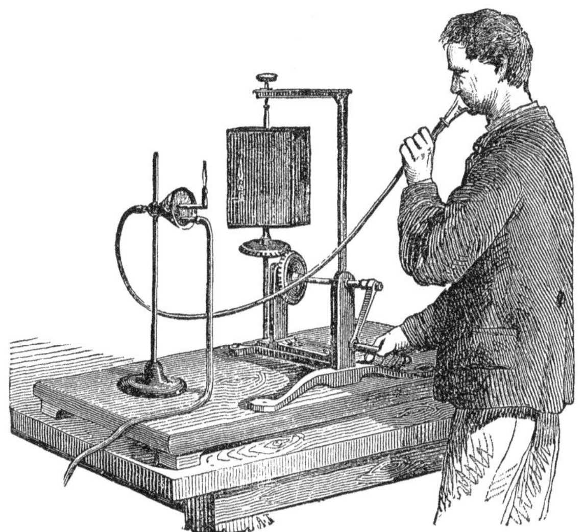

Fig. 97.—Vowels observed by the aid of Kœnig's Flames.

Kœnig obtained a visible image of the timbre of vowels by means of his flames, upon which the voice was made to act by a gutta-percha tube furnished with a funnel (Fig. 97). They are fed by a jet of gas, which crosses a hollow capsule closed on one side with a membrane, which is made to vibrate by the voice. This membrane acts upon the flame as a bellows, which makes it by turns flare up and grow dim; if the shocks be too violent, and the flame small, it is extinguished altogether; even if it be able to resist, it becomes

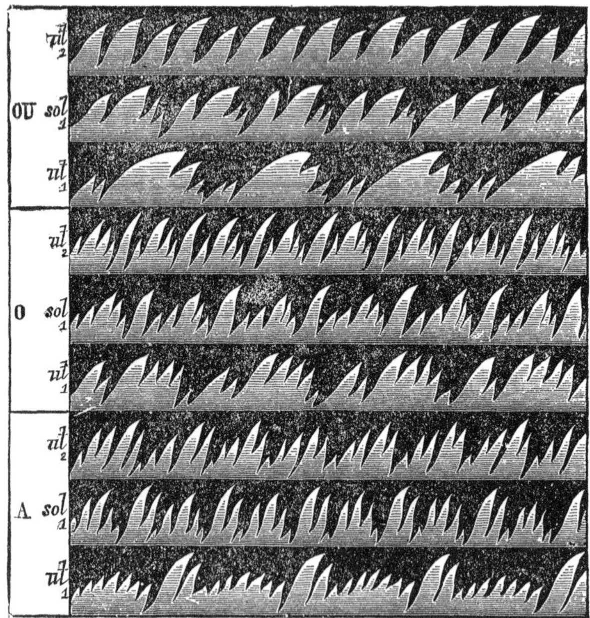

Fig. 98.—The Timbre of Vowels.

bluish. A flame palpitating thus would appear in the revolving mirror under the form of a serrated ribbon, whose changing appearance reveals the number and relative strength of its harmonics, as shown in Fig. 98.

After having accomplished the analysis of timbres,

Helmholtz tried to reproduce them by means of synthesis, reuniting the notes that had been separated by analysis. He constructed an harmonic series of eight tuning-forks, which were mounted between the branches of a set of electro-magnets, so as to be able to maintain them in vibration by the action of a periodical current of electricity. In front of each tuning-fork was placed a sounding-box, which could be shut more or less completely by pressing upon the key-board. When the box was closed the tuning-fork gave hardly any perceptible sound, but it grew stronger as the box was opened wider. With this apparatus, an *o* was distinctly produced by strongly sounding the si♭$_3$, more feebly si♭$_2$, and fa$_4$; *a* was obtained by giving si♭$_2$, si♭$_3$, and fa$_4$ moderately, and si♭$_4$ and re$_5$ with full force. The tuning-fork having si♭$_2$ for its fundamental gives, when sounding alone, a very faint *ou*. Kœnig made a like apparatus with ten tuning-forks. But we must always remember that compared with true vowels the resemblance is generally somewhat doubtful. Once, and only once, we heard a perfect *a*.

How is it possible to describe the marvellous power possessed by the ear for separating such complex sounds into simple vibrations? We have seen that the strings of a piano effect this dissociation of harmonics, since they answer to all the notes which are united together in the sound examined. Imagine a series of musical strings giving the scale of all possible notes, and then we shall have something with which to reproduce faithfully all varieties of timbre or composite sounds.

Helmholtz thinks that the ear possesses just such a series. This is the wonderful organ discovered by Corti, and called after him. It is situated in the *labyrinth*, and may be described as the terminal fibres of the auditory nerve. There are above 3,000 fibres spread over the membrane of this labyrinth; and, supposing that each one answers to a particular note, we have an instrument of 3,000 strings—more than enough to gather and reunite all the sounds in creation. There must be at least 400 for each octave.

In the same manner the perception of colours may be explained by the existence of fibres of the optic nerve, each appropriated to a simple colour. This hypothesis was put forward by Thomas Young. It cannot be denied that by this ingenious theory all the phenomena of our perception of colour and sound are explained in a very natural way. It is now understood that the ear must act as a prism, which decomposes the timbre into its primary elements, although the complex impression made upon the brain is seldom analysed by the mind accustomed to judge of its impressions only as a whole.

The most pleasant and musical qualities of timbre are the harmonics 1, 2, 3, 4, 5, 6. Compared with simple sounds, musical sounds are richer, fuller, and more magnificent—more *coloured*, so to say; they seem soft and mellow, too, so long as the sharp upper notes do not trouble the harmony. In this list we may place the sounds of the piano and organ, the human voice and the cornet, unless they are forced. The flute belongs rather to simple sounds. With such sounds only very little music can be produced: they must be sustained by others. An instrument composed of tuning-forks (which also give sounds almost simple) would not be pleasant to listen to alone.

The large pipes of an organ give very faint harmonics of the fundamental sound, and are therefore very nearly simple. This is especially true of the closed pipes. When a sound only contains the odd notes of the harmonic series (the fundamental, the twelfth, &c.), as happens in the narrow and closed organ-pipes, the clarionet, and strings

Fig. 99.—Voices of Birds.

struck in the middle, the timbre becomes *hollow;* when the number of higher sounds increases, it becomes *nasal*; when the fundamental sound governs, it is *full;* when this is too feeble, it becomes *thin.* The sound of a string is fuller when struck with a hammer than when pulled by the fingers.

When the harmonics above 6 are very distinct, the sound becomes harsh and piercing, because of the discords caused by these high notes; but if they are heard in moderation, they rather add brilliancy and colour to the tone of an instrument.

This subtle and changeful element which we have spoken of under the name of timbre plays an important part in the relations of voice and feeling. It is the timbre which renders a voice sympathetic, persuasive, and loving; or sharp, quarrelsome, and disagreeable. The timbre of a bird's song serves instead of speech, expressing all the emotions that stir its little heart (Fig. 99).

CHAPTER XII.

INTERFERENCE OF SOUND.

Beats—Resultant Sounds—Sonometers of Scheibler and Kœnig—Influence of the Movement of the Source of Sound on its Pitch.

HOWEVER paradoxical it may appear, we shall see in this chapter how sounds quarrel, fight, and when they are of equal strength destroy one another, and give place to silence. The phenomena of resonance revealed a sort of sympathetic reciprocal bond existing between sounds. The strings of a violin hanging on a wall resound without being touched when another violin is tried in the same room. Every sonorous body harbours a family of notes, which readily respond to the call of a friend. We are now about to study the warfare of notes, to spy out their enmities and discords. We shall see how the whole crowd of harmonics take sides when two declare war. Often, indeed, we hear them skirmishing when as yet the two chiefs are quiet.

Two notes are said to "beat" when their union gives rise to periodical alternations of strength and weakness. This phenomenon is well known in organ-pipes. When two slightly discordant pipes are sounded together there is a beating effect produced; the sound alternately swells and dies away, and when the strong swells follow very quickly there is quite a little tumult.

Sauveur was also the first to study this curious phenomenon, and he found that important deductions might be made from it. From his experiments he had concluded that the number of beats is always equal to the difference of height of the two notes; for each double vibration that the one accomplishes more than the other there is a beat; therefore nothing is easier than to determine the absolute height of two notes by counting their beats. Suppose, for instance, that two pipes are tuned for the notes doh and re: the interval being a major tone, the first will always make eight vibrations while the other makes nine; the difference being one, there will always be one beat for eight vibrations of one and nine of the other. If we now count four beats a second, we shall conclude that in the second the first tube has made four times eight, or thirty-two vibrations, and the other four times nine, or thirty-six; and thus the absolute pitch is at once determined. The beats may be also observed in tuning-forks, or in any other sonorous bodies, if only their vibrations be sufficiently slow.

What is the producing cause of this phenomenon of beats? According to Sauveur, "the sound of two pipes must have more power when their vibrations, after having been separated, reunite and coincide, and strike simultaneously upon the ear." "It even seems," he says, "that the common expression of musicians, that the pipes *beat* when their sound is thus redoubled, originates in this idea."

The explanation of beats rests upon the phenomena of interference. Two vibrations are said to interfere when they urge the air-particles in opposite directions. This is a case of "union is strength;" for when two vibratory motions, acting on a point, coincide, they assist and strengthen each other; when they are in opposition they weaken, and even annul the sound of both. In the same way it has been demonstrated that light added to light will produce darkness.

We have already seen the composition of vibratory motions, and how they may be examined by means of curves. Let us imagine two identical vibrations starting from the same point at the same time: they pass on uniformly, and acting together assist one another, and augment the motion of the particles, the result being a vibration in the same time,

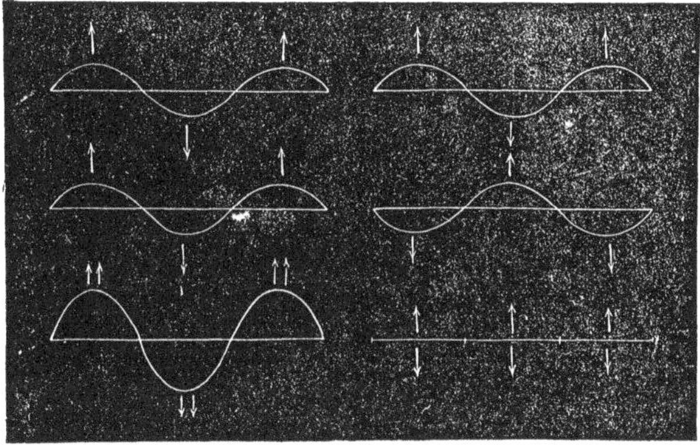

Fig. 100.—Coincidence. Fig. 101.—Opposition.

but much more energetic (Fig. 100). If the two vibrations be so disposed that one generates a condensation where the other generates a rarefaction, they act against one another, and if their power be equal, completely neutralise one another (Fig. 101). Two sounds of the same pitch and intensity thus meeting produce silence. This startling effect may be shown with two organ-pipes, exactly similar in all respects, and mounted side by side on the same bellows. While one only is played upon it sounds loudly; when both are made to sound together there is scarcely any sound, although they vibrate, as may be proved by placing a feather near the opening, where the current of air is broken; but they vibrate in opposition. When the air on entering one tube is condensed, it is rarefied in the other; the surrounding air is also urged in contrary directions by the two different actions; and since there is no reason why it should obey one rather than another, it remains motionless, and the sound is never produced.

This curious fact may be directly proved. A communication is made between the two pipes with two of Kœnig's

Fig. 102.—Interference.

flames, so arranged that the point of one passes a little mirror which hides its base, but shows by reflection the base of the other. This produces the illusion of a single flame. If, now, this flame be seen in the revolving mirror while the two pipes are played upon, the point will separate from the base, which proves that the two flames shine alternately (Fig. 102). If both pipes act on the same flame, the effect is neutralised, and the flame remains motionless. Two equal vibrations, then, either strengthen or weaken one another, according to the manner in which they combine; but the same effect, whichever it be, continues throughout the movement. If there be the slightest inequality the case is very different. In such a case one soon gains on the other, and passes on, then slackens, and is in its turn overtaken and passed, and so on. The encounters will take place in all manner of ways. Sometimes there will be an augmentation, sometimes a falling off, of sound; the two notes alternating more or less completely between brilliancy and extinction. If one should make exactly nine vibrations while the other made eight, and if the two vibrations started in opposition, they would at first weaken one another; then as one took the lead (nine simple vibrations having been accomplished on one side to eight on the other), they would coincide for an instant, thus supporting one another; then, after eight or nine simple vibrations more, they would again be in opposition, and weakened as at first. In the interval of eight and nine double vibrations, there would always be an augmentation of power or a *beat*. This would occur each time that the more rapid note gained a double vibration on the other (Fig. 103).

An illustration will explain this. Let us imagine two

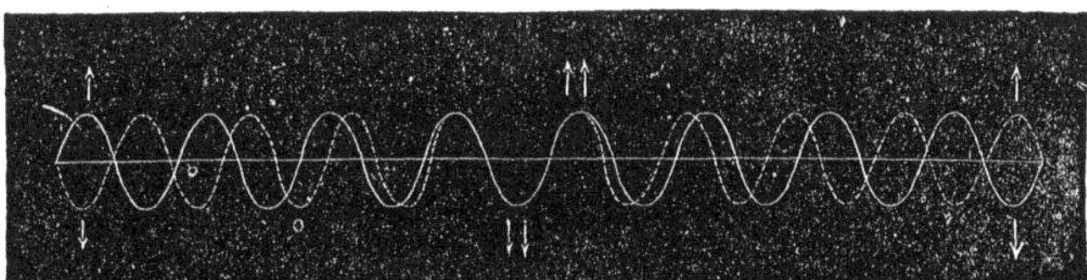

Fig. 103.—Beats.

rivers subject to periodical high-tides, rising in one at the beginning of each month, that is to say twelve times a year, and in the other every twenty-eight days, or thirteen times a year. Suppose further that between the high tides there intervened low ones: a high and a low tide would be equivalent to a complete undulation or double vibration. If these two rivers flowed into the same lake they must cause great commotion at certain times, whilst at others they would exert scarcely any influence over the state of the waters. It is indeed clear that if at any given moment the full tides coincide, the low tides must also fall together; and since the difference between the two rivers is but two days, this must happen for two or three months; and at these times their united action on the open waters would make a sensible rise and fall. But when the high tide of one happened at the same time as the low tide of the other, there would be no variation in the level of the lake. This period of calm would also last some months. Say that the two rivers rise together January 1st, they fall the 14th or 15th, mount again towards the end of the month, and fall in the middle of February. At the end of six months the river which rises every four weeks will be about fifteen days in advance of the other, and therefore will have a full tide in the middle of July, just as the other has a low one. This state of things will begin about June, and last till August. During this time there will be no effect. The summer then will be a period of great calm for the lake. Towards the end of the year, the second river being a whole month in advance, its thirteenth tide will coincide with the twelfth of the other, and the lake will again be agitated by a great flux and reflux. Thus each winter the lake will be stormy, and each summer will find it calm. In ten years, the 120 high tides of the one, combined with the 130 of the other, would have produced ten periods of maximum agitation. It is thus that two notes, making respectively 120 and 130 complete vibrations in a second, will give at the same time ten beats.

This phenomenon may be exhibited in various ways. By transcribing faithfully the vibrations of the air, the varying intensity will also be revealed if there have been any beats. To obtain a tracing with two slightly discordant tuning-forks, it is only necessary to fasten on one a piece of blackened glass, and on the other a flexible point; then make them vibrate horizontally, and hold them so that the point rests on the glass (Fig. 104). The curve then drawn

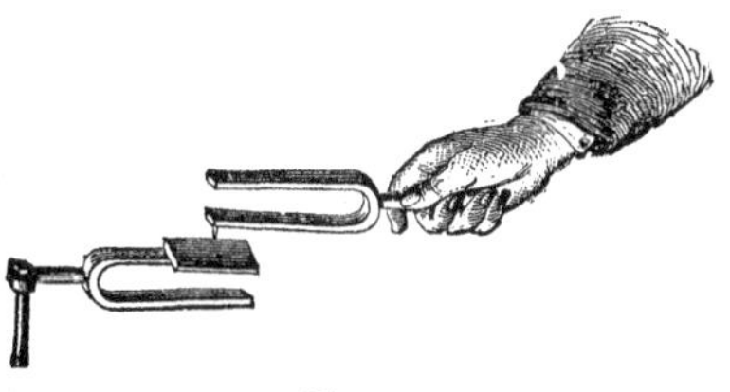

Fig. 104.

shows the augmentation as often as the one fork has gained on the other a complete vibration. Fig. 105 shows two tracings obtained in this way with two notes, which were at first in the ratio 24 : 25, and afterwards in that of 80 : 81. The flames of Kœnig furnish another means for observing beats.

The physiological perception of beats seems, at first sight, irreconcilable with the hypothesis, according to which the ear always separates notes of unequal pitch. If the two sounds do not act upon the same fibre, how can their vibrations combine in the auditory apparatus? The answer is simple. It must not be forgotten that the nervous fibres, like all elastic bodies, are influenced, though in a less degree, by vibrations a little out of unison, so that the sphere of action of two neighbouring sounds spreads over a large surface of fibres, instead of embracing only two. A note that is a semitone higher or lower than the note of a given fibre, makes it resound ten times less than a note in unison; still the resonance is perceptible. According to that, we see that the unison of two neighbouring notes, which *beat*, must be manifested in all the intermediate fibres, and the ear must be affected by them.

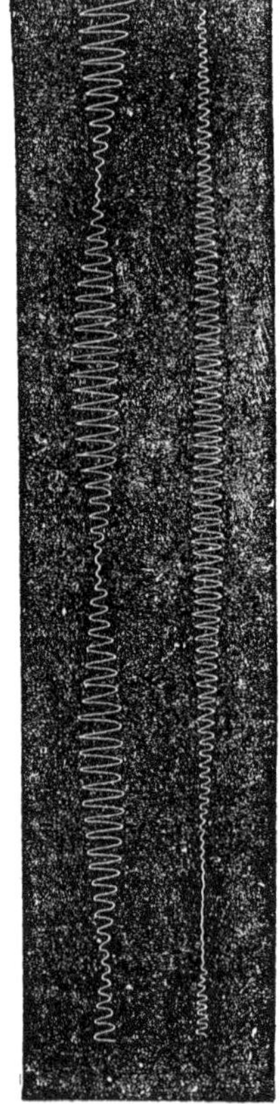
Fig. 105.

When the augmentations follow rapidly, the effect of the beats becomes very disagreeable, like the burr of an *r*, or the grating of a scythe on wood. The harshness is at its height when there are thirty or forty beats per second; beyond that it becomes difficult for the ear to separate them, and the impression is not so strong. Helmholtz declares that he has been able to distinguish up to 132 beats per second (between the si_5 and doh_6)—without counting them, be it understood. Since the lowest sound perceptible by the ear comprises about thirty double vibrations, it is therefore possible to hear beats at least four times as rapid as the lowest notes.

This observation contradicts the common opinion, that very rapid beats are perceived by the ear as a very low note. The reason of this hypothesis was, that two notes resounding forcibly together engender a third note, called the resultant tone, which is expressed simply by the difference of the two primitive notes, or, which comes to the same thing, by the beats produced by their concurrence.

The resultant sounds were known before they were understood. The German organist Sorge speaks of them in a work published in 1745. The celebrated violinist Tartini set himself, nine years later, to found a new musical system thereupon; but his book is so abstruse that even D'Alembert admits he could not understand it.

It has long been thought that the resultant sounds must always be lower than the sounds which cause them; but Helmholtz foretold by theory resultant sounds which should be sharper, and experiment has fulfilled his prophecy.

There are, then, two kinds of resultant sounds: first, the differential sounds, whose pitch is given by the difference in the number of vibrations of the primary sounds. These are the easiest to observe. Secondly, the additive sounds, the pitch of which is found by adding the vibrations of the primitive sounds. Let us suppose, for example, that two pipes are sounded together, giving a fifth. Their notes will be in the ratio of 2 : 3, and the difference being unity, the differential sound will be one, the octave below the lower of the two sounds. The sum of two and three is five; one might, therefore, also hear a note which would be the major sixth of the sharper of the two sounds. With doh_2 and sol_2 we can obtain doh_1 and mi_3, but we shall hardly hear anything beyond the doh—unless, indeed, the generating sounds are very strong. If (as generally happens) the latter are accompanied by harmonics, the intermingling of the respective harmonics, the fundamental notes, and the first resultant sound, may give birth to new resultant sounds; but these superfluities are difficult to observe, on account of their weakness.

The resultant sounds of a major third are these. The minims represent the primary, the crotchet the first differential, the quavers the cross products, and the barred note the additive sounds :—

To hear the resultant sounds, it is only necessary to force the generating sounds. Theory shows that this phenomenon must be considered as a kind of disturbance of the vibratory motion, which becomes too violent to follow the simple laws of ordinary elastic vibrations. It is by an analogous perturbation that tuning-forks and bells give the upper octave of their fundamental sound whenever they are violently set in motion; whilst vibrating moderately they would only produce upper sounds not harmonic.

Resultant sounds and beats render important aid in tuning organ-pipes, &c., indicating with great precision the difference in the pitch of two notes. Kœnig was thus enabled to tune a doh_9 of 32,000 vibrations, and a re_9 of 36,000, by their differential sound — the doh_6 of 4,000 vibrations.

Henri Scheibler, a silk manufacturer at Créfeld, did much to utilise the employment of beats for tuning musical instruments. This man, who had a passion for acoustics, devoted not less than twenty-five years to perfecting his method. He constructed, with inconceivable trouble, considering the state of science at that time, a set of fifty-six tuning-forks, giving the scale from la of 440 to la of 880, embracing an entire octave by degrees of eight simple vibrations. This set of forks formed what he called a sonometer. Taken two and two in the order in which they succeed one another, they always give four beats a second. They are thus tuned by differences, and the last will, of course, give the exact octave of the first. If this result has been attained we are sure that the first made 440, and the last 880 vibrations per second, for the beats prove a difference of 440, and we know, on the other hand, that they are as 1 : 2. We understand that these fifty-six tuning-forks, the notes of which are perfectly certain, allow any note whatever, contained within the limits of their octave, to be tuned by them with mathematical precision; we have only to count the beats that this note gives with the tuning-fork to which it stands in the nearest relation. If the note be in another octave it is set by means of a supplementary tuning-fork, which gives its true octave.

Scheibler published his method in 1834. He also went to Paris, to try and make his sonometer known; but the difficulty of construction frightened the manufacturers. Thanks to the progress of science, this valuable method is now within reach of every one. Kœnig made sonometers of sixty-five tuning-forks, embracing the middle octave of the piano (from 512 to 1,024 simple vibrations). He even went beyond this, filling in the same way the whole scale of perceptible sounds. In the bass octaves great forks are used, furnished with movable weights that slide along the branches; according to their position the fork gives different notes. In the very high octaves Kœnig replaces the tuning-forks by straight rods. The sonometer that he exhibited in 1867 was composed, first, of eight large tuning-forks, for the four octaves comprised between the doh of 32 and that of 512 simple vibrations. Each of these could give thirty-two notes, so that they represented a scale of 256 notes. Secondly, the middle octave (512 to 1,024) is represented by sixty-four; the next octave (1,024 to 2,048) by eighty-six; and the next (2,048 to 4,096) by 172 tuning-forks, making a total of 330. Thirdly, from doh_6 of 4,096 vibrations, Kœnig employed steel rods, the length of which is inversely proportional to the pitch of their longitudinal sound. Ninety-six rods thus represent the four octaves from doh_6 to doh_{10} (64,000). This last octave is almost beyond the limit of perceptible sound; few people can hear the sol_9 (48,000) that Kœnig obtained by transverse vibrations of a rod about three inches long.

Two tuning-forks with an exact difference of two simple vibrations will beat the seconds just like a pendulum; if they vary more, they will beat a fraction of a second as small as is wished for. In counting these beats we may also see another very curious phenomenon—the influence of a movement of the sonorous source on the pitch of its note. Kœnig took two tuning-forks, doh_4, giving four beats per second when left in their places; he placed himself about two feet distant from the sharper one, and moved the other backward and forward between it and his ear, keeping his eyes fixed on a pendulum. When the to-and-fro movement was synchronous with that of the pendulum, the listener only heard three beats in the second when the low tuning-fork approached his ear; but there were five when it receded. It follows that the tone of this fork was raised a double vibration during the first second, and was equally lowered during the following one. In fact, by moving it two feet nearer (which represents the length of its wave) a complete vibration is gained, and by moving it an equal distance the same is lost—just as navigators who sail round the world gain or lose a day, accordingly as they travel with the sun, or in a contrary direction.

Railways often afford opportunities for observations of this kind. Thus the whistle of the engine-driver seems more shrill when the train approaches than when it is passing

Fig. 106.—Influence of Motion on the Pitch of Sounds.

away. Taking thirty-one miles an hour as the speed of a train, we find that it moves about forty-six feet per second, which is $\frac{1}{24}$ of the velocity of sound; a calculation based upon this shows that for an observer placed on the railroad, the note of the whistle will be changed in the ratio of 24 : 25; he will either estimate it too high or too low by a semitone, according to the direction of the motion. If it is a la for the engine-driver, it will be la♯ for the signal-man at the approach of the train, and la♭ after it has passed. A stationary whistle would have the same effect for passengers;

they would only hear the true note at the moment of passing. If the observer and the whistle were carried in opposite directions, the effect would be still more striking—the note would appear alternately a whole tone higher, and a whole tone lower than the reality. At the moment the trains met it would leap a major third.

In 1845, M. Buys-Ballot made some experiments of this kind on the railroad between Utrecht and Maarsen. Three groups of musicians were placed as close as possible to the rails, and distant from one another about half a mile. A musician placed upon the locomotive blew a trumpet, first on leaving Utrecht, then between the three groups, and finally after having passed them. The others estimated the varying pitch of the note, and it was always found conformable to theory.

Mr. Scott Russell tells us that the reflection of the noises of a train on the piles of a bridge should produce the same effect as the contrary movement of two trains, and thus the notes which are echoed back, altered by a whole tone, mix very discordantly with those which are heard directly. To obtain minor thirds by reflection, the train should move at a speed of seventy-three miles an hour.

A German philosopher, named Doppler, has inquired into these facts, applying them to luminous vibrations, and the explanation of the colours of the stars; but these are only speculations.

CHAPTER XIII.

THE VOICE.

Organ of the Voice—Bass—Tenor—Alto—Soprano—Celebrated Voices—Song and Speech—Vowels and Consonants—Ventriloquism.

THE sublime effects of the human voice are produced by a very puny instrument. Some cartilages, a pair of ligaments, a group of muscles—that is all which Nature needed to create a musical instrument, the sweetness and moving power of which no human invention has rivalled. This vocal apparatus is a reed with two lips. It is composed of the *larynx*, a cartilaginous tube, which forms the "Adam's apple" in the throat; the *vocal chords*, flexible ligaments with only a narrow slit, the opening of the glottis, between them; the *lungs*, which furnish the wind; and the *cavities of the mouth*, where the first rude sound of the voice is fashioned into vowels and consonants.

The vocal chords can meet and separate, contract and expand, by the action of certain muscles; the current of air proceeding from the lungs makes them vibrate, and this vibration causes the sound. Thanks to that ingenious instrument, the laryngoscope, by means of which the inside of the mouth is made visible and the formation of the voice may be observed, the different conditions which modify it are well known.

For the production of a chest-voice a very complete action is necessary, and a very close contact of the two sides of the glottis ; and the vocal chords vibrate throughout their whole extent. In falsetto notes they only vibrate partially, and the glottis opens so as to form an elliptical orifice. Practised singers can sound the same note alternately in chest-tone and falsetto, without taking breath ; but as for Garcia's story of the Russian peasants, who sang an air simultaneously with chest-voice and falsetto, we must class it among the miracles.

If the voices of women be shriller than those of men, it is because of the smaller dimensions of the larynx. The opening of the glottis is nearly twice as large with men as with women and children. At the age of puberty the glottis of a man suddenly enlarges, and his voice generally drops an octave ; it is then said to "break."

Men's voices are divided into bass, barytone, tenor, and counter-tenor. The last-mentioned is at the present day extremely rare. Women's voices are contralto, mezzo-soprano, and soprano. In the following table is shown the compass usually assigned to these different voices.

This shows that ordinary voices do not compass two full octaves. The difference between the lower fa in the bass (174 simple vibrations) and the upper sol of the soprano (1,566 vibrations) is a little over three octaves. But these limits

The bony partition opposite the tympanum is perforated by two small holes, the one round and the other oval; they are both closed by fine membranes. The oval orifice, which is above the other, communicates with the tympanic membrane by a series of little bones. These are—the *hammer* (*m*, Fig. 108), which is fastened to the middle of the tympanum; the *anvil* (*n*), resembling a molar tooth, and supporting the head of the hammer; the little lenticular bone (*l*) joining the *anvil* to the *stirrup-bone*, which adheres by its base to the membrane of the oval orifice. Some tiny muscles attached to the sides of the concha can act upon the hammer and anvil, making them turn together round a horizontal axis; the end of the hammer then either draws or pushes the tympanic membrane, and the end of the anvil acts on the stirrup.

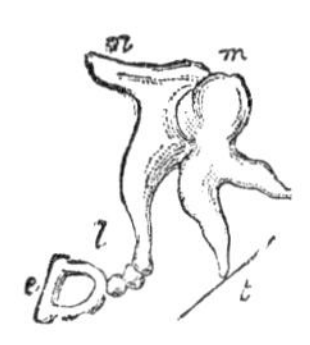

Fig. 108.—Ossicles.

The internal ear, or labyrinth, is composed of the vestibule V (Fig. 107), surmounted by three semi-circular canals R, and the cochlea L, which has the form, both outside and inside, of a turbinated shell. The vestibule opens on the oval orifice, the cochlea on the round one; but they communicate by means of a pretty large opening. The bony labyrinth has a lining membrane, which takes nearly the same form, and is generally a counterpart of the external surfaces. It is filled with a liquid called the vitreous fluid, and over it are spread the terminations of the acoustic nerve N.

The process of hearing is as follows:—The vibrations of the tympanum are communicated by the air in the concha to the round orifice, and by the ossicles to the oval orifice. The membranes which close these orifices make

several passages that she sang before him. We only quote the last, which ends in doh$_6$.

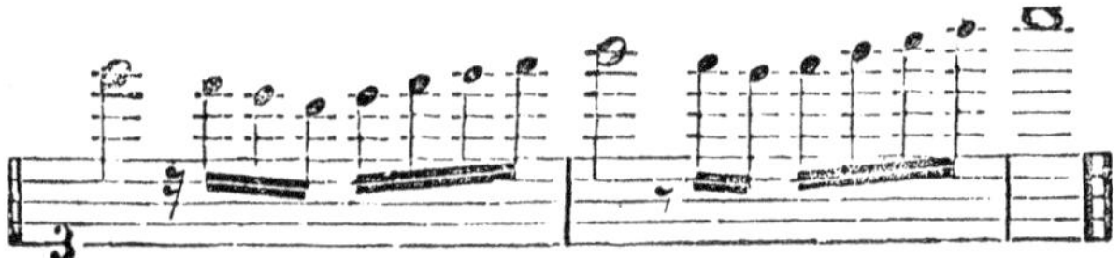

Trills were given on the re$_6$, and other adornments of a similar kind. The father of Mozart adds that La Bastardella sang these passages with a little less power than the lower notes, but her voice remained pure as a flute. She descended easily as low as sol.

Oulibicheff tells of a Madame Becker, who astonished St. Petersburg in 1823 by her wonderful roulades. Kuhlau composed the part of Adelaide, in his opera *Le Château des Brigands*, for her. The grand air in the third act goes up to la$_5$. On one occasion, at the moment of giving this dangerous note, the leader of the orchestra looked so fixedly at her that she was frightened, and gave doh$_6$.

The quality of the voice depends, as before explained, on the number and force of its harmonics. A "true voice" is one that passes without hesitation from one note to another. Practice will do much to produce it, but a musical memory is also necessary. The absolute pitch of the notes is difficult to fix in the memory; but it is by no means uncommon to find people, especially professional musicians, who can give any note as it is asked for by name.

The difference between the singing and the speaking voice consists in this: the first bounds from interval to interval, while the conversational voice rises and falls by a continuous motion. The singing voice is sustained on the same tone, as on an indivisible point, which is not the case in simple pronunciation, where the sounds are not sufficiently united to be appreciated from a musical point of view.

The dramatic declamation of the ancients was an approach to song, and often had an accompaniment on the lyre. We find a relic of it in the peculiar intonation of the Italian orators, and in the monotone recitation heard in cathedrals. Recitative forms the link in modern music between speech and song. It might even be said that up to a certain point song is but an idealised imitation of the accents of impassioned speech. One may cry and complain without singing, but both may be imitated by song. With a little attention, too, we may find the vestiges of musical intonation in common speech. The accented syllables and the fall of phrases are marked by a change of tone. In an affirmative German sentence, Helmholtz says, the point is indicated by a fall of a fourth, while in an interrogation it rises a fifth. Indications of this kind are to be found in the Gregorian chant.

In Chinese, intonation is a grammatical element.

"If," said M. Ch. Beauquier, "we could translate into musical sounds all the most *singing* sentences, such as interrogations, menaces, ironical sayings, &c., we should find a national similarity, amongst different individuals, in accenting the same phrases. The Italian modulates much, the German less, the Englishman not at all."

The sounds of speech are divided into vowels and consonants; the timbre of the vowels varies according to the resonance of the mouth, but the consonants are, as we have explained already, little more than noises. The lips, the tongue, the palate, the teeth, bear a part in the production of these characteristic sounds, which make up the framework or scaffolding of speech, and which are alone written by the Orientals, to the utter neglect of the vowels. The child commences with the vowels, only gradually learning the consonants, and only when he does so can his speech become intelligible.

The letters of the alphabet have been thought by some to have certain physiological characters. Listen to Mersenne. He writes as follows:—

"The vowels *a* and *o* signify what is grand and full; and because *a* is pronounced with a widely opened mouth, it signifies clear things, and actions which are used in opening or beginning some work. Therefore it was that Virgil commenced his 'Æneid' by the word *Arma.*

"The vowel *e* expresses something subtle, and is properly used in mourning and sorrow:

'Heu quæ miserum tellus, quæ me æquora possunt!'

"The vowel *i* means very small and slight things. Thence comes the word *minim.* It expresses also something penetrating.

"*O* is expressive of strong passions: *O patria! O tem-*

pora! *O mores!* and to represent rotundity, because the mouth must form a circle while uttering it.

"*U* belongs to things secret and hidden."

Then he proceeds to classify the consonants. He makes the *f* indicate a breath, a wind (*flatus*); *s* and *x*, bitter things (*stridor*); *r*, rough, hard, disorderly things, violent and impetuous actions, which have earned it the title of the canine letter; *m*, all that is great (*magnus*, *monstre*); *n*, things dark, hidden, and obscure, and so on.

Boiste, in his "Observations on Pronunciation," says the *e* is the soul of the French language; it is the most variable of all the letters, the one most capable of modulation, and having most shades. According to the same author, "the doubling of the *f* denotes either sharpness or vanity, pedantry or satire; it irritates, it domineers, it bites, in such words as *en effet*, *qu'ai-je affaire*, *cela suffit*, *c'est affreux*. None but a born and educated Frenchman can truly pronounce this." Words formed by onomatopœia imitate natural noises; the great poets often get very happy effects from the different characters of consonants and vowels. In the well-known verse of Virgil the clatter of horses' hoofs is rendered by a succession of vigorous dactyls:

> "Quadrupedante putrem sonitu quatit ungula campum."

It has been remarked that each of the vowels has its favourite place in the musical scale. Helmholtz says that the vowels which belong to a given note are, first, those whose characteristic is a little higher than the note in question, and afterwards those whose characteristic is the octave or twelfth of the same note. The *ou*, whose characteristic is fa_2, is produced with greatest ease on the notes re_2, mi$_2$, fa$_2$, and fa. The *e* prefers re$_3$, mi$_3$, fa$_3$: then again fa$_3$ and si♭, because of its characteristic fa$_3$.

This affinity of the vowels for certain fixed notes is principally verified in the limits of falsetto and chest-voices. A woman's voice giving a lower note than doh$_3$ turns involuntarily to the *o* or *ou*. Above mi$_4$, the most easy note to sound is *a*. Passing si$_4$, *i* takes the ruling place. Such facts are very important to composers, and to those who write words intended for a musical setting.

Jean Müller and other physiologists have studied the mechanism of the human voice, by means of the artificial larynx, made of india-rubber bands fixed to the end of a tube, and acted upon by pincers, which give a variable tension. By blowing into this apparatus, sounds can be produced closely resembling those of the human voice.

To imitate the vowels, the theory of timbre shows that it is necessary to strengthen certain fixed notes in these sounds. Thus it is that Mr. Willis produces the vowels by the help of a whistle mounted on a tube, which he could lengthen or shorten at pleasure. By adding to such an apparatus sensitive membranes to produce the characteristic sounds of consonants, it is possible to imitate speech. We have all heard of the dolls who say *papa* and *mamma*. Mr. Wheatstone had a kind of bag-bipe which could pronounce short phrases. Mersenne tells us of an organ that gave vowels and consonants. In 1791 Van Kempelen exhibited a speaking automaton, but the spectators did not speak very enthusiastically of the resemblance of the artificial sounds to the human voice.

The vocal apparatus found in birds is placed very low in the throat. This is the reason that Cuvier was able to cut the neck of a singing bird without preventing its song. With men an accidental opening in the larynx renders the formation of the voice impossible. Magendie tells us of a man he knew who was always obliged to wear a cravat with a valve, to stop a leakage in his throat.

The organ-stop called *vox humana* is only a set of very short zinc pipes, which often give a harsh and screaming sound, and is seldom effective.

Ventriloquists only talk like ordinary mortals, but they avoid opening the mouth so as to be *seen* to speak, and they scarcely move their lips, and breathe as little as possible. Their voice then appears changed, and as if coming from a great distance. This is not done without a great effort of the lungs, which fatigues the chest, and obliges the ventriloquist from time to time to resume his natural voice; therefore dialogue is easy to them, while at the same time it helps to mislead the audience. They speak also while breathing, and the stifled sound thus produced seems to come through a thick wall. The illusion is completed by an imitation of the inflexions used when people call from a distance. But when one becomes familiar with the voice of a ventriloquist the illusion is dispelled. Robertson proved this with a servant of his, who was a famous ventriloquist.

Ventriloquists generally find it very easy to imitate the voice of a child; but they can rarely sing in a borrowed voice.

This art was known in the earliest ages; the sorcerers

made use of it. Amongst the celebrated ventriloquists we may mention Louis Brabant, valet-de-chambre of Francis I., Saint Gille, Baron van Mengen, Charles, Comte, &c. Of this last they tell a number of odd stories. Once at Tours he made them break open a closed shop, in which, from the groans they heard, they supposed some one was starving. At Nevers an ass suddenly declared, with strong invectives, that he would carry his rider no further. He cured people possessed with devils, exorcising the demons, who were heard to fly away howling. In a church invaded by revolutionists greedy of destruction, he made the statues speak, reproaching the iconoclasts for their Vandalism; and they took to flight, wild with terror. Once he saved himself from the peasants of Fribourg, who were going to burn him as a sorcerer, by making a voice of thunder come from the furnace towards which they led him, whereupon they fled in disorder.

CHAPTER XIV.

THE EAR.

The External and Internal Ear—The Ossicles—The Mechanism of Hearing—The Fibres of Corti—Inequality of the Two Ears—Perception of the Direction of Sound.

ON either side of the head Nature has placed the ears, commissioning them to receive and introduce to the presence of the mind the sounds which arrive as invisible messengers from Nature. It is not because there is no other way in which the auditory nerve can be reached. We have seen that it is possible to hear through the teeth. Deaf people have even been known to hear by the epigastrium; but the natural road for sonorous impressions is by the auditory canal.

With men and all the mammalia, the hearing organ comprises three successive compartments—the external orifice, the middle ear, and the inner ear. The external ear is composed of a passage C (Fig. 107), opening out at the base of the temporal bone, and a cartilaginous funnel. This is a sort of hearing-trumpet, for gathering and concentrating the sonorous waves. When this is wanting, or is flattened against the head, the hearing loses much of its delicacy. In many animals this concha is movable; horses and dogs prick their ears, in order to hear better. The motion is produced by the cutaneous muscle of the head. It is a very rare faculty among men, though individuals possessing it are occasionally met with.

The middle ear is separated from the external by the tympanic membrane, which closes a kind of hollow cavity in the hardest part of the temporal bone. This membrane receives the sonorous vibrations, and transmits them to the interior. With birds and reptiles it is placed near the crown of the head. The passage E forms a free communication between the membrane and the back of the mouth,

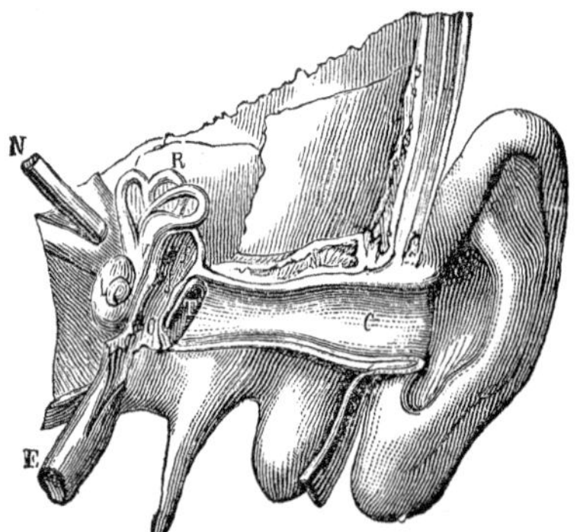

Fig. 107.—The Ear.

by which means an equilibrium is maintained between the outer air and that imprisoned in the cavity. One can easily experience the reality of this communication, by stopping the mouth and the nose, and then blowing. The tympanic membrane swells under the pressure of the imprisoned air, and one who tries to breathe under these conditions feels it to be drawn inwards. This explains why we should open the mouth when standing near a cannon as it is fired. The pressure upon the tympanum caused by the detonation is thus diminished, by being equalised on both sides.

are passed by exceptional voices. On the one hand we hear of bass voices reaching the fa of 87 vibrations; and on the other, of sopranos that can touch fa in the fifth octave of 2,784 vibrations, and even higher.

The voice of Gaspard Forster, a Dane, extended over three octaves, while that of the youngest of the sisters Sessi embraced three and a half. Catalani could also command three octaves and a half.

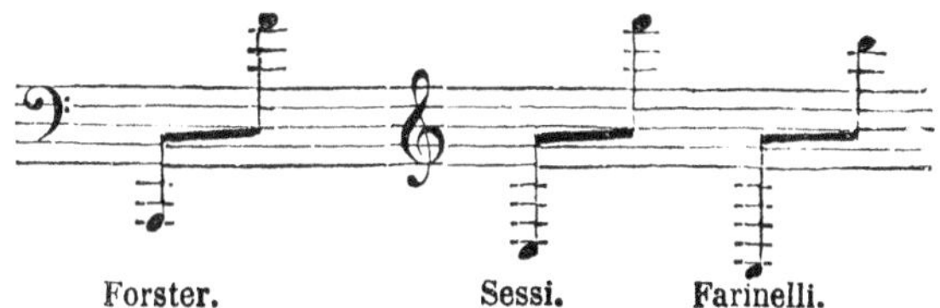

At the Bavarian Court there were in the sixteenth century three remarkable basses, who, according to Prætorius in his "Syntagma Musicum," reached the fa_{-1}.

Christine Nilsson and Carlotta Patti attain a marvellous height. When acting the Queen of Night in *The Magic Flute*, Mdlle. Nilsson gives the fa_5. But the highest voice ever known seems to have been that of Lucrezia Ajugari, whom Mozart heard in Parma, 1770. In a letter addressed to his sister Marianne, he transcribes

the fluid in the labyrinth vibrate, and consequently the floating filaments of the acoustic nerve; and thus the sensation of sound is produced.

The hammer, probably, serves also to give a variable tension to the tympanic membrane when we listen attentively. The movement of the muscles which control it may be voluntary. Fabrice d'Aquapendente could produce a little noise in his ear by acting on the *hammer*, and Müller could make his *ossicles* crack, so as to be heard by another person. M. Daguin observed, when he was handling some very small objects in perfect silence, and let one fall by accident, a slight tinkling, due in all probability to the same cause. These facts prove that the hammer strains the tympanic membrane when one "gives ear," just as the pupil adjusts itself to look fixedly at an object.

The tympanum is not absolutely necessary to hearing. When it is torn the hearing is impaired, but not destroyed, since the surrounding air then acts directly upon the membranes of the two orifices.

The inner membrane of the cochlea is lined with elastic fibres, discovered by the illustrious Corti, and bearing his name. They apparently form the terminations of the filaments of the auditory nerve. Helmholtz thinks that each one is attuned to a special note, and as they are above 3,000 in number, there must be above 400 for each octave. The interval from one to another would be $\frac{1}{66}$ of a tone, and so they form a wondrous instrument for reproducing every note that the ear can distinguish. We have already seen its bearing on timbre, and the analysis of harmonics. The cochlea may, then, be called an Æolian harp of 3,000 strings, that move in sympathy to all the sounds of creation.

This idea has been unexpectedly confirmed by the

recent researches of M. V. Hensen on the hearing of the decapod crustaceans. Having placed some of these animals in sea-water, charged with strychnine, in order to intensify the action of the nervous centres, he has seen them thrown into convulsions at the slightest noise; from which experiment he concludes that with them audition takes place through the medium of auditory hairs, each hair vibrating in unison with a certain note. When he examined the point of attachment between a nervous cord and one of these hairs under the microscope, whilst a horn was being loudly sounded, the point became indistinct through the rapid motion of the hair each time certain notes were given, the neighbouring hairs remaining motionless. One of the hairs answered to re$\sharp_2$ and to re$\sharp_3$, a little more faintly to sol$_2$, and still less to sol. Probably it had for its fundamental tone an harmonic common to these four notes, and situated between re$_4$ and re$\sharp_4$. Another hair vibrated under the influence of the notes la$\sharp_2$, re$\sharp_2$, and la$\sharp$, which indicated the fundamental tone la$\sharp_3$.

In the vestibule and semi-circular canals, the terminations of the nerves are found to be under other conditions. There we find some little crystalline particles, called otolithes, and some fine elastic bristles, which seem meant to sustain the vibrations of the nervous filaments. Scarpa and Treviranus believed this different formation of the various ramifications of the acoustic nerve must be for the purpose of enabling us to distinguish the pitch and timbre of sounds; but our present knowledge of the matter does not allow us to define everything in the wonderful organisation of the auditory apparatus.

Paralysis of the auditory nerve causes incurable deafness. Atrophy of certain parts of the plexus, or "Corti's organ," explains the partial deafness which prevents sounds of a certain pitch being heard. Many ears are incapable of hearing very high sounds. Wollaston found that several people were deaf to the chirping of crickets, and some even to that of sparrows. Why may there not be animals to whom sounds beyond the range of the human ear are still perceptible? There are certain kinds of insects that vibrate, just like our well-known crickets, without making the faintest audible sound: may it not be that there is, in truth, a delicate music audible only to its proper listeners?

Musicians have been known to play in the orchestra, and to be aware of the slightest falsity in tune, who yet could not join in a common conversation without the help of an ear-trumpet. Mr. Willis describes a singular phenomenon under the name of *paracousis*. Some people who have imperfect hearing, and cannot in general hear faint sounds at all, hear them at once when they are accompanied by a great noise. Mr. Willis knew a lady who made her servant beat a drum whenever she wanted to listen to anything, for then she could hear very well. Another person could only hear when the bells were ringing. Holder mentions two similar cases: one of a man who was deaf except when they beat a large drum close beside him, and another of one who never heard so well as when he was rattling over a stony road in a carriage. There was a shoemaker's apprentice, too, who only heard while his master was beating the leather on the stone. Such facts may perhaps be explained by the habitual relaxation of the muscles of the *hammer*, which would render them incapable of acting on the tympanic membrane, except under the excitement of very strong vibrations.

With many people the ears are unequal in their power

of hearing. From M. Fechner's experiments it seems that the left ear generally hears better than the right. He thinks that the reason may be the common habit of sleeping on the right side. Ittard mentions a remarkable instance of a man he knew, whose two ears heard different notes at the same time when a single one was given. M. Fessel, of Cologne, lately discovered the same peculiarity in himself. In setting some tuning-forks—first by his ear, and then by a more exact process—he noticed that all which he had set by his right ear, while holding the normal or pattern tuning-fork to his left, were too low, while the others, set in the opposite way, were too sharp. It follows that the same sound is sharper for his right ear than for his left. Much struck by this circumstance, he examined the hearing of many persons, and found it a much more common thing than could have been imagined. So that we might ask a musician for the right la or the left. M. Fessel even supposes that the phenomenon is objective, and that the same tuning-fork really gives a higher note when it vibrates before the ear to which it appears sharper. This note of resonance is heard in the same way by an on-looker. He asked different persons of his acquaintance to carry alternately to the right and left ear two duplicate tuning-forks; and according to the notes they heard, he could tell by which ear they heard too high or too low. But such facts need verification.

As the two eyes serve to give us the impression of the geometrical relief of a body, so the two ears allow us to judge of the direction of sounds. When the eyes are blind-folded, and one ear stopped, all sound seems to come in the direction of the free ear; or, at least, our inference as to its direction is very uncertain.

It is the concha of the ear that specially helps to arrest attention, and to recognise the direction of sonorous waves. Diderot tells us of a blind man who, when disputing with his brother, took up something and threw it very neatly at his head, his aim being guided by his ear.

The hearing of the blind is generally very acute and delicate, because it has to serve them for the most part instead of sight. Ittard invented an instrument, which he called an *acoumeter*, to measure the delicacy of hearing. It is a brass ring, hung upon a thread, and struck by the ball of a pendulum which falls from a given height. The distance at which different persons cease to hear it is accurately measured. Freycinet used this instrument for studying the hearing of savages. In nocturnal birds and timid animals, such as the hare, the external ear is largely developed. The ears of the lower animals are incomplete. The cavity of the tympanum is entirely wanting in fish, the round and oval orifices being at the top of the head. The articulata do not show any visible auditory apparatus. Amongst the molluscs, the cephalopods are the only creatures that possess a vestige, and there it is of the simplest form, consisting merely of a cavity and acoustic nerve.

CHAPTER XV.

MUSIC AND SCIENCE.

Principles of Music—Euler—Rameau — Sauveur — Helmholtz — Harmony and Discord Explained by Beats—Chords—Major and Minor Keys.

THE disdain with which the majority of musicians reject all attempts of the exact sciences to invade their domain is, up to a certain point, justifiable.

The help that mathematics has hitherto given to musical science is very slight; it has scarcely done more than point out a few vague analogies which explain nothing. It has travelled in a defective circle; the pleasure of the ear has been exalted into a principle, and made the foundation of all systems.

It was known that harmonious chords correspond to the relations of whole numbers. The Pythagoreans propounded and repropounded this theory, without deducing from it any other conclusion than some aphorisms upon the harmony of the world, and the occult power of numbers. Philosophers have attempted to find the seven notes of the scale repeated in the movements of the celestial bodies, and even the great Kepler abandoned himself to such mystical speculations.

In the first half of the eighteenth century, towards 1740, the great mathematician, Leonard Euler, endeavoured to explain the relations of musical intervals by considerations drawn from physiology. He reasoned thus: That which pleases us is always that which to our feeling possesses a certain perfection; and wherever there is perfection there is necessarily also order—that is to say, some law which governs. A song will please us if we recognise the order of the sounds of which it is composed; and it will please us so much the more in proportion as we are able to understand that order. Now, there are in sounds two ways in which order may manifest itself—by their pitch, as represented by high notes or low notes, and by their duration. Pitch is reckoned by rapidity of the vibrations, and duration by the length of time during which a sound is heard. Order with regard to duration consists in rhythm or time; order in pitch is simple proportion amongst the vibrations. The degrees of accord in these proportions—that is to say, in the musical intervals—depend upon their simplicity, for the ear appreciates them so much the more easily as they are expressed by the most simple numbers, and the pleasure is greatest when it costs us least. In developing these principles Euler succeeded in establishing the laws of harmony.

That which is wanting in his theory is, that it is not based upon any certain fact. Nothing warrants us in admitting that the ear can judge of the relations of vibrations which depend on the thousandth part of a second. The observations of astronomers show that the ear separates at the most two strokes of a pendulum which vibrates in a tenth parth of a second. How can it be supposed that it can compare the proportion between two vibrations numbering, for example, 5,000 and 5,050? And nevertheless it easily recognises this relation in so many musical intervals.

Ideas analogous to these of Euler had been already put forward in 1701 by Sauveur. "The mind," he says, "by its very nature loves, at the same time, simple perceptions because they do not weary it, and varied perceptions because they spare it the ennui of uniformity. . . . . Every variety which pleases the mind is then confined in certain limits; it must be guarded from becoming difficult to perceive, confused, complicated . . ." He then explains how chords are rendered agreeable to the ear by the more or less frequent concurrence of vibrations. When these concurrences become rare, as in thirds where they occur only once in five or six vibrations, the perception of the sound becomes less simple, but it is nevertheless pleasant because it is slightly varied, the discords putting the harmonies in still stronger contrast.

But there is a point at which the harmony of this variety stops, and this point is given by the ratio 5 : 6. Sauveur afterwards remarks that harmonies do not make beats, and that discords do. Unfortunately, he has not developed this idea as it deserves.

In 1726, Rameau started another theory, which D'Alembert thought worthy of notice. It seems, at first, to account for the pleasure that music gives us. It is very curious to see the means that this celebrated artist has taken to discover what he calls *the principle of harmony.*

"I saw," said he, "that I must follow in my researches the same order that exists in the things themselves; and as, to all appearance, there must have been song before there was harmony, I asked myself in the first place how song was obtained.

"Enlightened by the method of Descartes, which I had happily read, and with which I had been struck, I began with myself. I tried some songs, like a child who is practising singing; I watched what took place in my mind and in my voice, and it always seemed to me that there was not any reason that decided me, when I had uttered a sound, to choose one more than another of all the multitude of sounds that might come next. There were certainly some for which my voice and my ear seemed to me to have a predilection, and that was the first thing I noticed; but this predilection appeared to me purely a matter of habit. I imagined that in a different system of music from ours, with another kind of song, the predilection of the voice and sense would have been in favour of another sound; and I concluded that, since I found in myself no good reason to justify this predilection and to regard it as natural, I must not take it as a principle in my researches, nor even suppose that it would exist in another man who was not in the habit of singing or of hearing it."

He declares, however, that the sounds which had seemed to him to succeed each other most naturally were the *fifths* and the *thirds*, or the sounds which correspond to the relations of 2 to 3, and of 4 to 5. But this simplicity of relations appeared to him to be only a sort of convenient arrangement, and insufficient to account for a phenomenon such as that which he sought to explain.

"I began," continued he, "to look around me, hoping to find in nature that which I could not discover in myself so surely or so clearly as I desired. The search was soon rewarded: the first sound which struck my ear was a clap of thunder. I suddenly perceived that it was not one, and that the impression which it made upon me was complicated. 'That, said I, is the difference between *noise* and *sound*. Everything that produces a simple impression upon my ear is *noise*, and everything that produces an impression composed of several others is *sound*.' I called the primitive sound a fundamental tone; its concomitants, harmonics."

He afterwards discovered that harmonics are very sharp

and very transient, so that they cannot strike equally a musical ear, and one lacking in musical sensibility. Then he decided that the complementary sounds of the fundamental tone must be its twelfth and its seventeenth—that is, the octave of the fifth and the double octave of the major third. Then, as he knew by experience, as he says, that the octave is only a *repeat*, he thought it quite natural that his voice and his imagination should lower the harmonics to the last point; and that, therefore, his fancy should be taken by the third and the fifth of the fundamental tone, and not by their repeats, when he took the notes that his ear suggested to him after the fundamental tone.

Thus the multiple resonance of the sonorous body becomes the base upon which is built the musical system. Rameau deduces from it the formation of the diatonic scale and the principal rules of harmony. But his fertile imagination led him afterwards to attempt to draw from the same source the principle of geometry; and it is here that D'Alembert, to whom is due the merit of developing and simplifying Rameau's system, felt himself obliged to place his veto, and to circumscribe clearly the range of the musician's discovery. D'Alembert continually asserts that the demonstration which Rameau pretends to have given of the principles of harmony is no demonstration, and that there will always enter into the theory of musical phenomena a sort of metaphysics, which introduces into the science an obscureness natural to itself. "But," says he, "if it be unjust to demand here the unshaken complete assurance which is produced only by the clearest light, we doubt at the same time whether it would be possible to throw upon these matters a stronger light than that which we have already."

The judgment that D'Alembert passes upon Rameau's

system proves sufficiently that the illustrious mathematician understood perfectly well its weak points, or, to speak more correctly, its insufficiency. It is not enough to say that the octave is a *repeat;* the word does not sufficiently account for the important part that this interval plays in musical compositions; and, on the other hand, the phenomenon of harmonic resonance is not so general as Rameau supposes. A large number of sonorous bodies produce in reality entirely dissonant simultaneous sounds. It is therefore not right to lay it down as a principle that harmonics are found by *natural* resonance; and even if it were true, we must remember that the ugly has quite as much a place in nature as the beautiful, which proves that a thing may be at the same time natural and disagreeable.

It must then be acknowledged that this theory has not a rational foundation, since it does not explain in any way the origin of discords. Nevertheless, we cannot but admire the ingenuity with which Rameau has deduced his system from data so incomplete; and it may be said, without exaggeration, that he has inaugurated a new era in the theory of music.

The celebrated Tartini published in 1754 a treatise on Harmony, in which he took as his starting-point *resultant tones,* which he thought he had discovered, and which he had observed when he played two chords at once. Tartini calls such tones of the series 1, 2, 3, &c., *monad harmonics,* from the concurrence of which results a sound. All harmony, he says, is comprised between the monad, or component unison, and the full sound, or compound unison. He then enumerates the resultant tones of musical intervals, always mistaking the octave, and finds that the different intervals may be so arranged as to give the same resultant tone, which may be considered, therefore, their common base, &c. &c.

At this time the theory of music had not emerged from a circle of ideas completely estranged from natural philosophy and physiology. Generally, the propounders of systems have lost themselves in mystical speculation. The German philosopher Herbart travelled in this track; according to his views any two sounds suggest to the mind two ideas, which exercise at once an attractive and repulsive force. In the soul of the fifth, hate has just overcome love; in the major third, the two powers keep an armed neutrality. The most curious conclusion is, that the adjusted scale is that which satisfies most fully the musical ear! and that Herbart was the first to lay the foundations of a mathematical psychology.

Aristoxenus had eagerly combated the arithmetical subtleties of the Pythagorean school. He has found many imitators among musicians of modern times. The Spaniard Eximeno published, towards the end of the last century, a work in which he demonstrates that music has no manner of connection with mathematics. This must still be the opinion of M. Fétis, to judge from the preface to his "Traité d'Harmonie."

This learned theorist describes in the following terms his discovery of the principles of harmony—a discovery which he made one day in May, 1831, as he was travelling from Passy to Paris, and which caused him such emotion that he was obliged to seat himself at the foot of a tree:—"Nature furnishes us, for elements of music, with only a multitude of sounds, which differ among themselves in intonation, duration, and intensity, in a greater or less degree.

"Amongst these sounds, those which differ sufficiently to

affect the sense of hearing distinctly become the objects of our attention; the idea of the relations which exist between them becomes present to the intelligence, and under the action of sensitiveness on the one hand, and will on the other, the mind arranges them in different series, of which each one corresponds to a particular order of emotions, of feelings and ideas.

"These series become then the types of tones and rhythms which have necessary consequences, under the influence of which the imagination comes into exercise, in the creation of the beautiful."

After such assertions, ought he not to have made the scale?

There appeared in Germany, in 1863, a book which made immediately a great sensation. It was "La Théorie de la Perception des Sons," by Helmholtz.* The illustrious author has succeeded in reducing to physical phenomena, susceptible of being submitted to calculation, the secret relations of sympathy and antipathy which exist between natural tones, and explaining the cause of the sensations which we experience from them.

M. Helmholtz is Professor of Physiology at the University of Heidelberg, which boasts also Kirchhoff and Bunsen. Already illustrious by the discoveries with which he has enriched physiological optics—it is to him that we owe the ophthalmoscope—and by other scientific researches, he was the man who was needed to find the answer to an enigma two thousand years old.

We have already spoken at length of the researches to which M. Helmholtz devoted himself, with the object of

* Die Lehre von den Tonempfindungen.

discovering the true nature of tone, and we have mentioned his experiments on beats and resultant tones. It was in that way that he discovered the key to harmony, the true principle of concords and discords.

It is necessary to fully understand his ingenious arguments, and with that view we first consider beats. The disagreeable sensation that they give us is easily explained. All intermittent excitement of the nerves tires us. We know the unpleasantness of unsteady light like that of a flame blown by the wind. A strong steady light soon dulls the irritability of the retina, just as continued pressure hardens the skin; a flickering light, on the contrary, or a rapid and oft-repeated pressure, allows the nerves to retain their sensibility, and becomes for that reason a source of pain. Tickling excites the epidermis in the same way an intermittent sound irritates the ear, and hence it is that beats are felt to be a source of discord.

Sauveur had divined the same reason. "Beats," he says, "do not please the ear, because of the inequality of the sound, and it is very probable that it is the absence of these beats which renders octaves so agreeable. Following out this idea, it appears that the chords in which the beats are not heard are just those that musicians call harmonies, and that those in which the beats are perceived are discords; also that when a chord is discord in a certain octave, and harmony in another, it is because it beats in one and not in the other; it is then called an imperfect concord. If this hypothesis be true it will reveal the true source of the rules of composition, till now unknown to philosophy, which referred almost entirely to the judgment of the ear. Natural judgments of this kind, however foolish they may sometimes appear, are not in reality so; they have some very real causes, the knowledge of which belongs to Philosophy, provided she were able to put herself in possession of it."

Sauveur (or rather Fontenelle, the historian of the Academy) adds afterwards, in returning to this idea, that what is real harmony of chords has probably not been fixed by Nature, and that what is called a fine ear is quite as much the result of long custom, of old habits, and of arbitrary prejudices, as of an inborn faculty. He would thus explain the great difference that is found between nations in their taste for music.

These ideas were not further developed, and they fell into oblivion. It is only recently that M. Helmholtz has entered upon the same investigation with all the resources of modern science, and has unravelled the physical principles of harmony.

In studying beats, M. Helmholtz first proves that the degree of roughness which they give to a musical interval does not depend solely upon their frequency; they become less irritating in the bass octaves, where the same number of beats correspond to a larger interval. Thus, the minor second, si_3—doh_4, is very discordant, while the fifth, doh—sol, is a harmony; and yet these two intervals give alike thirty-three beats in a second. This circumstance is explained by the greater difference of the strings, which answer to a larger interval. The sol does not vibrate the string allotted to doh, and the doh does not vibrate the string sol, whence it follows that resonance is powerless to unite the two notes on the same string, and to give rise to beats; on the contrary, the notes si and doh make a great number of strings vibrate in common, which renders their beats perceptible to the acoustic nerve.

When the beats are observed in two tones between

which the interval is very great, the phenomenon is due to the harmonics, or rather to the resultant tones. Thus doh$_2$, the harmonic of doh, will beat with all the notes which it may happen to approach; for instance, with re$_2$ or si, even when these notes occur as harmonics of another fundamental tone. Two tones, too far removed to touch each other directly, may then be in opposition through the medium of their satellites; thus, mi$_3$, the harmonic of doh, will beat with mi$\flat_3$ which carries the colours of la$\flat$. But the struggle may even take place under the same roof; when two harmonics of the same note find themselves too close together they quarrel. Thus, the harmonics 8 and 9, or 9 and 10, which differ only by a tone, always beat, and disturb the internal harmony of the tone, wherever they are at all prominent. Their presence explains the harshness of the trumpet, and of strained bass voices.

When two sounds of whatever tone make exactly the octave, the harmonics of the sharper note are superposed upon the harmonics of the flat.

Doh ...	1	2	3	4	5	6	7	8	9	10	. . .
Doh$_2$ ...		2		4		6		8		10	. . .
	doh	doh$_2$	sol$_2$	doh$_3$	mi$_3$	sol$\sharp_3$	la$_3$	doh$_4$	re$_4$	mi$_4$	. . .

From that time there are no more beats; but however little the chord may be disturbed, we become aware of it by the great tumult that the divided harmonics produce. The doh$_2$ will beat with the untrue doh$_2$, the doh$_3$ with the untrue doh$_3$, and so on.

The reason why the octave is the consonant interval *par excellence* of which the ear has the most correct appreciation is easily seen. The virtual or eventual beats of the harmonics distinguish it by their energy, the least discord betraying itself by a great cacophony. The other concords are much less characteristic, as we shall see. Take, for instance, the twelfth 1 : 3, and the following will be the order of the two series :—

Doh...	1	2	3	4	5	6	7	8	9 . . .
			|			|			|
Sol ...			3			6			9 . . .
	doh	doh_2	sol_2	doh_3	mi_3	sol_3	. . .	doh_4	re_4 . . .

The coincidence of the harmonics again takes place here; but it is less important. If the doh be a little untrue, the harmonics 3, 6, 9, which it has in common with the sol_2, are divided and beat; but they are weaker than the harmonics of a less elevated order, which divide when the interval of the octave is adjusted; their beats are not so perceptible, and then the concord is less distinct.

The other concords—fifths, fourths, thirds, &c.—contain already elements of discord; here the harmonics are superposed only partially, but there remains the germ of discord. Thus, for example, in the fifth :—

Doh ...	2		4	6	8		10	12 . . .
				|				|
Sol ...		3		6		9		12 . . .
	doh	sol	doh_2	sol_2	doh_3	re_3	mi_3	sol_3 . . .

The sol_2 and the sol_3 are, at the same time, the harmonics of doh and sol, and coincident when the fifth is exact; but the re_3 of the series of SOL can beat with the doh_3 and the mi_3 of the series of DOH. The concord of the fifth is then not absolutely pure, besides which it is less characteristic than the octave; for a false fifth only makes those harmonics beat which are of the same class as those that beat a false twelfth.

The same may be said of the other harmonious chords. The more slightly raised the harmonics are which form the coincidence, the purer is the interval, and the better distinguished by the eventual beats of these harmonics.

In the intervals where there exist harmonics which have the power to disturb the chord, it is necessary to take account of the juxtaposition more or less close of these notes, for the beats will be so much the more slow in proportion to their nearness. We have already said that the impression made by thirty-three beats in the second is very disagreeable; beats much more rapid than this cease to be perceptible; and very slow beats, instead of annoying the ear, give to the music a solemn character, or an expression of trembling emotion like that produced by the tremulo of the voice.* It follows that an interval will be so much the more discordant, as it supplies a larger number of less elevated harmonics which are able to produce *beats of a certain rapidity.*

On these principles it is easy to calculate *a priori* the degree of purity of different intervals considered in all parts of the musical scale. M. Helmholtz calls an interval in which one of the two given notes coincides with a partial tone of the other, an absolute or free consonance, for in that case there is also coincidence between all the respective harmonics. To this category belong unison, successive octaves, the twelfth, seventeenth, &c. The intervals which immediately follow in point of purity are, first the fifth, then the

* There is found in modern organs, in fact, a regular arrangement for making beats. The effect of the register called *unda maris* is made also by slow beats.

fourth, which may be called perfect concords; the major sixth and third are medium concords; the minor sixth and third are but imperfect ones. Yet the beats of the thirds are very sensible in the deep notes of the scale, and they were not admitted into the group of imperfect concords until the end of the twelfth century. The employment of minor sixths and thirds is never justified except by necessities in the formation of chords.

If the intervals be split, the major fifth and third are improved (they change into the major tenth and twelfth); the fourth, the minor third, and the sixth, on the contrary, become more discordant.

M. Helmholtz tried to make these phenomena, and the laws which regulate them, evident by means of a figure, which represents in a very irregular curve the relative degree of discord or any two notes of a violin, calculated according to the intensity and frequency of the beats of the superior tones of those notes, supposing that the highest effect would be thirty-three beats in a second. Upon a straight line, by which is represented a note which, starting from doh$_3$, rises by insensible degrees to the double octave doh$_5$, stand the Cor-

Fig. 109.

B

dillera of displeasure (Fig. 109). Valleys mark the position of unison of the fifth, the octave, the twelfth, and the double octave. The Chimborazo of discord occurs quite close to unison, where the least discord produces the most perceptible beats. More or less distinct unevenness distinguishes the other discordant regions ; and more or less deep depressions, the various degrees of concord.

The influence of resultant tones is in every way analogous to that of superior or harmonic tones. The union, then, of two tones accompanied by their harmonics, the first differential sounds, produces only the beats pointed out as being those of harmonics; and, as they are in general much more feeble than harmonics, the consideration of them is less important for practical purposes, where we have to do only with musical tones that have harmonics; but in treating of simple tones, it is necessary to have recourse to the beats of resultant tones, to account for discords, and to characterise harmonics. Thus the first differential sound of the octave coincides with the deepest of two given notes, and can therefore beat with it, since the chord is disturbed; and this is one means of judging of the accuracy of an octave formed by two simple notes. The fifth again, and perhaps also the fourth, are characterised by resultant tones; but the other intervals lose all clearness and decision when only simple tones are employed. And this is, in fact, the reason why empty harmonic tones are improper in musical harmony: they can only be used to strengthen richer tones. This remark applies, for instance, to the large closed pipes of an organ. If a piece of music be played upon an organ with the register closed, it has neither character nor energy; the absence of harmonics makes it very difficult to distinguish harmonies from discords, and this want of clearness renders the music so weak and soft as to be tedious.

The sound of the flute contains, besides the fundamental tone, its sharpened octave, and sometimes the twelfth; the intervals of the octave and the fifth are well defined; the thirds and sixths only very indistinctly. It is a common saying that the worst thing in the world after a flute solo is a duet on two flutes; yet this instrument becomes very useful when it is played in concert with others which have more energy. The same thing may be said of harmoniums with diapasons. Therefore we see that the quality of musical intervals varies necessarily with the tone of the instruments.

The most extended analysis of the sound of instruments has shown that the ear delights, above all, in tones in which the two first harmonics (octave and twelfth) are strongly accentuated, the two following somewhat modified, and the others less and less perceptible. Taking this as a starting-point, it is easy to explain the particular effect of each instrument, and to establish *a priori* a number of practical rules known to musicians.

It is clear that the consideration of beats helps to the understanding of the part that whole numbers play in the fixing of musical intervals. Fourier's law, in virtue of which every sonorous movement is an accumulation of simple notes, becomes thus the true base of counterpoint, since concords are derived from the superposition of partial sounds, and discords from their antagonism.

We have now to speak of sounds with respect to their effect produced when they are combined in music; this subject encroaches upon the domain of æsthetics, where we have no longer fixed and invariable principles to guide us like those of purely physical sciences. Musical scales, modes, &c., have been developed, step by step, in the course of centuries; and the changes that the tastes of different nations have wrought in them are a sufficient proof of the instability of their foundations. The science of counterpoint is based, in part at least, upon laws capable of improvement, and it would be rash to affirm that it has yet reached its last point of development. Nevertheless, here again we find some general laws which seem to have guided artists unknown to themselves, and which spring naturally from those which we have already established. These laws prove the philosophical necessity of rules to which ignorant groping has led.

Thus the formation of multiple chords rests upon the same principles as that of consonant intervals. It is necessary that the three intervals between the three notes which compose a triple chord should be separately consonant, in order that the chord may be so. Intervals which exist in different chords may be classified under different degrees of consonance.

The difference between major and minor modes may consist in the resultant tones which are formed by the combination of three notes. In major chords the resultant tones are only repetitions of notes given in lower octaves. It is found that in minor chords this does not happen; the resultant tones there are formed by the harmony, and form major chords which accompany the minor. This intervention of a strange element, and probably also the very feeble beats of resultant tones of the second order, give to the minor chord something mysterious and undecided, that all musicians have felt without being able to account for.

In the accompanying example the major and minor chords are printed in minims; the resultant tones of fundamental notes, in crotchets; the resultant tones due to the combination of fundamental notes and harmonics, by quavers and semiquavers. A rest placed after a note denotes that it is slightly higher than the sound it ought to represent.

Passing on to the melodious combination of sounds, we find that melody depends like harmony upon the phenomenon of superior tones, inasmuch as these tones determine the affinity of sounds, just as the affinity of chords results from the notes which are common to them. Melody is the succession of notes following each other in an order pleasant to the ear. According to Rameau and D'Alembert, it springs from harmony, and the effect of it will be found expressed or unexpressed in the harmony, and especially in the unexpressed fundamental bass. But as homophonous song existed long before polyphonous music, or music in harmony, we are compelled to seek an independent origin for melody.

We notice first that melody is a movement which is produced by a change in the height of notes, and which we can conceive imitated by mechanical movements. But the mind would not have been able to appreciate, or even to feel, these shades of expression if the progression of the notes had not been arranged according to a definite value —that is, by intervals of tones or half-tones, and in a fixed rhythm.

The *bar* helps us to divide time; the progression of notes by tones or semitones allows us to separate the height of notes into fractional parts; and thus we understand movement by rhythm and melody. The sensations that we experience at the sight of a rough sea, when the waves follow each other at regular intervals, are of the same nature. In the voice of the wind the notes blend without intermission, therefore they produce upon us a painful and confused impression, through the absence of all proportion and distinctness. Music, on the contrary, has a standard for measuring the ascending and descending movement of tones, and this standard is the scale.

But why were the notes of which the scale is composed adopted? Was there a reason for it? Why do we find there the octave, with its fifths, fourths, and thirds? The answer is easy, after what we have already remarked concerning partial tones or harmonics. The following table represents the harmonics with consonant intervals:—

Tonic	(1)	1	2	3	4	5	6	7	8	9
Octave	(2)	—	2	—	4	—	6	—	8	—
Twelfth	(3)	—	—	3	—	—	6	—	—	9
Fifth	($\frac{3}{2}$)	—	—	3	—	—	6	—	—	9
Fourth	($\frac{4}{3}$)	—	—	—	4	—	—	—	8	—
Third	($\frac{5}{4}$)	—	—	—	—	5	—	—	—	—
Third	($\frac{6}{5}$)	—	—	—	—	—	6	—	—	—

The octave with its attendant harmonics being comprised in the tone of the voice, it is clear that in ascending the octave a fractional part of the tonic is constantly repeated. Therefore we may say, with Rameau, that the sharpened octave simply answers to the tonic, the harmonics of which, 2, 4, 6 . . . . it reproduces. It is in this sense that the successive octaves of a key-board are only repetitions of the same scale.

The twelfth, being the third partial tone of the tonic, is equally expressed by the tonic, but less completely than the octave, because it only produces the harmonics 3, 6 . . . of the tonic. Lowering it an octave, we have the fifth, of which the second partial tone reproduces the harmonic 3 of the tonic, the fourth the harmonic 6 of the tonic, and so forth. The fifth is then, again, a partial echo of the tonic ; but at the same time it contains new notes which are not comprised in it, and has therefore less affinity for the tonic than the octave and the twelfth. The affinity of the fourth is still less, for there it is only the third partial sound, which corresponds with the fourth of the tonic. Therefore it was that the polyphonous songs of the Middle Ages were accompanied by fifths. The thirds and sixths answer to the tonic still less perceptibly ; they were introduced into music only at the time when true harmony began to develop itself.

M. Helmholtz calls those tones which have at least

one harmonic in common, affinities of the first degree; and two sounds which have an harmonic in common with a third sound, he calls affinities of the second degree. Building upon this foundation, he succeeded in constructing in a very reasonable manner a diatonic scale of notes, which have either the first or second degree of affinity for the tonic.

The direct relatives of the tonic doh are composed of the notes doh_2, sol, fa, la, mi, and mi♭, if we stop at the first six harmonics, the others being too weak to determine the affinity. We have then the scales—

Doh — — mi — fa — sol —la — —doh_2;

or—

Doh — — mi♭ — fa — sol — la — — doh_2;

for two notes so similar as mi and mi♭ could not be introduced into the same scale.

In order to divide the two excessive intervals which exist in this series, it is necessary to have recourse to the relatives of sol, which consist of the notes doh, re, mi♭, si, doh_2. The re and the si are united to doh by an affinity of the second degree, and by inserting them into the scales given above, the diatonic scale

Doh — re — mi — fa — sol — la — si — doh_2

is obtained, which becomes the minor ascending scale if we put mi♭ in the place of mi. The re which would be taken amongst the relatives of fa, would differ by a comma from the re fixed by sol. These examples suffice to render the method followed by M. Helmholtz comprehensible.

In studying the rules of harmony, it becomes evident that chords, considered as complex sounds, have amongst them the same relations of affinity as the notes of the scale, in consequence of the coincidence of some of their notes. The importance of the tonic in modern music, or that which M. Fétis calls the principle of *tonality*, is also explained by the nature of the superior tones of the tonic. These clear and simple principles have allowed of the fundamental rules of composition being deduced from mathematical considerations, which M. Helmholtz has done. Nevertheless, it must be confessed that the theory of music is not yet completed; all the deductions that M. Helmholtz has drawn cannot be considered fully proved, and they are not universally admitted. For instance, M. Arthur von Oettingen has criticised (and with reason) the explanation that M. Helmholtz gives of the difference between major and minor, for the phenomenon of harmonics is sometimes very little apparent. M. d'Oettingen traces this difference to the reciprocal principles of *tonics* and *phonics*.

The tonicity of an interval or of a chord consists in the possibility of considering it as a group of harmonics having the same fundamental tone. Thus, the major chord is formed of the harmonics 4, 5, 6 of the tonic, or fundamental bass, 1. The phonicity of that interval would be the inverse property of having an harmonic in common; the minor chord $\frac{1}{6}$, $\frac{1}{5}$, $\frac{1}{4}$ has the tone 1 as its common harmonic or phonic. The major chord has for its phonic 60; the minor chord has for its tonic $\frac{1}{60}$. The relations may be explained as follows:

$\frac{1}{60}$	— $\frac{1}{6}$ - $\frac{1}{5}$ - $\frac{1}{4}$ —				4 - 5 - 6 —	60
Tonic	— Minor chord —	Phonic	1	Tonic	— Major chord —	Phonic
fa	— la-doh-mi —	mi		doh	— doh-mi-sol —	si

Musicians call doh the tonic, and sol the dominant, of

the scale of doh major, which may be written in this way :—

Doh	re	mi	fa	sol	la	si	doh
1	$\frac{9}{8}$	$\frac{5}{4}$	$\frac{4}{3}$	$\frac{3}{2}$	$\frac{5}{3}$	$\frac{15}{8}$	2

M. d'Oettingen calls mi the phonic, and la the leading note, of la minor; and writes this scale in the following manner :—

Mi	fa	sol	la	si	doh	re	mi
$\frac{1}{2}$	$\frac{8}{15}$	$\frac{3}{5}$	$\frac{2}{3}$	$\frac{3}{4}$	$\frac{4}{5}$	$\frac{8}{9}$	1

In developing this dualism he establishes the parallel construction of the major and minor modes. But we must draw to a close details which have, perhaps, already wearied the reader.

If it be possible thus to establish *a priori* the most important laws of music, however grand may be the result with regard to the philosophy of the art, it does not follow that the knowledge of these laws is all that is required in a musician. We must here repeat what D'Alembert has said in the preface to his book on music: "Nature must do the rest; without her, no one will compose better music for having read these elements, any more than he would write good verses for possessing Richelet's Dictionary. In a word, it is the elements of music that I pretend to give, and not the elements of genius."

In the works of art that we admire, we instinctively divine a secret law which the artist has obeyed, however ignorantly, and it is in this sense that we must use the words of Leibnitz so often quoted: *Musica est exercitium arithmeticæ occultum nescientis se numerare animi*

When the law is so manifest that it instantly strikes the eye, we feel the intention and the calculation, and the work does not move us; for one essential condition of admiration is, not to understand completely. Admiration ceases as soon as we feel ourselves on an equality with the artist. This is the unconscious law which distinguishes a work of art from a systematic and calculated production; it must not therefore be supposed that science can, or ought to, discover and lay bare all the resources of the creative intellect.

THE END.

An Important Historical Series.

EPOCHS OF MODERN HISTORY.

EDITED BY

EDWARD E. MORRIS, M.A., and J. SURTEES PHILLPOTTS, B.C.L.

Each 1 vol, 16mo, with Outline Maps. Price per volume, in cloth, $1.00.

HISTORIES of countries are rapidly becoming so numerous that it is almost impossible for the most industrious student to keep pace with them. Such works are, of course, still less likely to be mastered by those of limited leisure. It is to meet the wants of this very numerous class of readers that the *Epochs of History* has been projected. The series will comprise a number of compact, handsomely printed manuals, prepared by thoroughly competent hands, each volume complete in itself, and sketching succinctly the most important epochs in the world's history, always making the history of a nation subordinate to this more general idea. No attempt will be made to recount all the events of any given period. The aim will be to bring out in the clearest light the salient incidents and features of each epoch. Special attention will be paid to the literature, manners, state of knowledge, and all those characteristics which exhibit the life of a people as well as the policy of their rulers during any period. To make the text more readily intelligible, outline maps will be given with each volume, and where this arrangement is desirable they will be distributed throughout the text so as to be more easy of reference. A series of works based upon this general plan can not fail to be widely useful in popularizing history as science has been popularized. Those who have been discouraged from attempting more ambitious works because of their magnitude, will naturally turn to these *Epochs of History* to get a general knowledge of any period; students may use them to great advantage in refreshing their memories and in keeping the true perspective of events, and in schools they will be of immense service as text books,—a point which shall be kept constantly in view in their preparation.

THE FOLLOWING VOLUMES ARE NOW READY:

THE ERA OF THE PROTESTANT REVOLUTION. By F. SEEBOHM, Author of "The Oxford Reformers—Colet, Erasmus, More," with an appendix by Prof. GEO. P. FISHER, of Yale College. Author of "HISTORY OF THE REFORMATION."

THE CRUSADES. By Rev. G. W. Cox, M.A., Author of the "History of Greece."

THE THIRTY YEARS' WAR, 1618-1648. By SAMUEL RAWSON GARDINER.

THE HOUSES OF LANCASTER AND YORK; with the CONQUEST and LOSS of FRANCE. By JAMES GAIRDNER of the Public Record Office.

THE FRENCH REVOLUTION AND FIRST EMPIRE: an Historical Sketch. By WILLIAM O'CONNOR MORRIS, with an appendix by Hon. ANDREW D. WHITE, President of Cornell University.

THE AGE OF ELIZABETH. By Rev. M. CREIGHTON, M.A.

THE FALL OF THE STUARTS AND WESTERN EUROPE FROM 1678 to 1697. By Rev. E. HALE, M.A.

THE PURITAN REVOLUTION 1603-1660. By S. R. GARDINER.

THE EARLY PLANTAGENETS. By WILLIAM STUBBS.

☞ *Copies sent post-paid, on receipt of price, by the Publishers,*

SCRIBNER, ARMSTRONG, & CO.,
743 and 745 Broadway, New York.

www.ingramcontent.com/pod-product-compliance
Lightning Source LLC
LaVergne TN
LVHW010236110826
845151LV00004B/1316

* 9 7 8 1 4 2 5 5 2 6 2 8 3 *